AN INTRODUCTION
TO
THE STUDY OF
EXPERIMENTAL
MEDICINE

BY
CLAUDE BERNARD

Translated by
Henry Copley Greene, A.M.

With an Introduction by
Lawrence J. Henderson

With a new Foreword by
I. Bernard Cohen
Professor, Harvard University

DOVER PUBLICATIONS, INC.
NEW YORK

This Dover edition, first published in 1957, is an
unabridged and unaltered republication of the first
English translation originally published in 1927 by
Macmillan & Co., Ltd. A new Foreword has been
specially written for this Dover edition by I. Bernard Cohen.

International Standard Book Number: 0-486-20400-6
Library of Congress Catalog Card Number: 57-3629

Manufactured in the United States of America
Dover Publications, Inc.
180 Varick Street
New York, N.Y. 10014

FOREWORD

The usual definition of a scientific "classic" is a great work that is venerated, cited, but no longer read. Claude Bernard's book is an exception in that it is still read as part of the philosophy program for the baccalauréat in French schools and is even now displayed prominently in booksellers' windows near the École de Médecine in Paris. Published in 1865, and issued in an English translation in 1927 (reprinted in 1949), *An Introduction to the Study of Experimental Medicine* is one of the few major works of Claude Bernard available in English.[1] The introduction to the English translation, written by the physiologist L. J. Henderson, has rightly acquired a fame of its own. This book was intended by Bernard to present the basic principles of scientific research, illustrated by case histories taken from his own work. At once the reader obtains a clear and penetrating view of the nature of science and an insight into the growth of the ideas of one of the greatest of all men of science.

Like all classics, Bernard's *Experimental Medicine* must be read with an awareness of both the general intellectual climate and the state of knowledge when it was written—in this case about a century ago. Although Bernard fully appreciated the importance of mathematics and said that "the application of mathematics to natural phenomena is the aim of all science," he believed that many of the attempts to apply mathematics to physiological problems were faulty because the empirical data were insufficient. He held, therefore, that "the most useful path for physiology and medicine to follow now is to seek to discover new facts instead of trying to reduce to equations the facts which science already possesses." He did not categorically condemn the application of mathematics to biological phenomena, but only insisted that "since a complete equation is impossible for the moment, qualitative must necessarily precede quantitative study of phenomena." Examples of the types of calculations of which he disapproved are given on pages 131ff. Anyone would agree with the absurdity of making a "balance sheet" of every substance taken in and excreted by a cat during eight days of nourishment and nineteen days of fasting, if on the seventeenth day kittens were born and

[1] *Illustrated Manual of Operative Surgery and Surgical Anatomy,* by MM. Cl. Bernard and Ch. Huette (New York and London: H. Baillière, 1852; reissued 1855);

counted as excreta (p. 132). Another example given by Bernard is the physiologist who "took urine from a railroad station urinal where people of all nations passed, and who believed he could thus present an analysis of *average* European urine!" (p. 135)

The present reader, however, will find greater difficulty in agreeing with Bernard's strictures on the use of statistics. Bernard simply could not understand "how we can teach practical and exact science on the basis of statistics . . . [which can] bring to birth only conjectural sciences," but which "can never produce active experimental sciences, i.e., sciences which regulate phenomena according to definite laws. By statistics, we get a conjecture of greater or less probability about a given case, but never any certainty, never any absolute determinism." Since "facts are never identical," (pages 138-139) statistics can serve only as "an empirical enumeration of observations." Hence if medicine were based on statistics, it could "never be anything but a conjectural science; only by basing itself on experimental determinism can it become a true science, i.e., a sure science." Here Bernard was expressing the difference between what he denominated the point of view of "so-called observing physicians" and that of "experimental physicians." He believed that medicine was in a "lowly state," perhaps even "wholly conjectural." The complexity of the phenomena in medicine made it more difficult for that subject to become an exact science than, say, physics or astronomy. But he refused to accept the "anti-scientific ideas" of his contemporary medical thinkers who held that "medicine cannot but be conjectural" and who therefore inferred "that physicians are artists who must make up for the indeterminism of particular cases by medical tact." These physicians resorted to statistics because they believed that in medicine and in physiology "laws are elastic and indefinite." But Bernard held that "if a phenomenon appears just once in a certain aspect, we are justified in holding that, in the same conditions, it must always appear in the same way. If, then, it differs in behavior, the conditions must be different." This situation is contrasted with "indeterminism [which] knows no laws; laws exist only in experimental determinism, and without laws there can be no science."

This rejection of statistics and the implied indeterminism of their application by medical men is closely linked with Bernard's endeavour to transform physiology into an exact science. Determinism is, in fact, the guiding principle in the thought of Claude Bernard, as it was for his teacher Magendie. It derived from the conviction accepted as a truism today, but still in question a century ago, that biological phenomena are as regular as chemical and physical phenomena, and subject to the same type of exact experimental laws. From today's point of view the deterministic aspect of chemical and physical phenomena holds only for events on a large scale, not on the microscopic—atomic and subatomic—scale. But even today the type of indeterminism necessary for micro-physics has found little place in experimental biology, and none whatever in the consideration of the gross phenomena of life with which Claude Bernard was concerned.

For the main part, the arguments against the vitalists have today only an historical interest, although the pungency of Bernard's attack remains a source of delight to the reader. The description of how a scientist attacks his problems continues, however, to be as splendid a statement of the basic features of scientific research as has ever been written, its authority strengthened by the importance of the scientific achievements of the author which are used as illustrations. Descartes, whom Bernard greatly admired, wrote a "Discourse on Method," which served as a general introduction to three scientific treatises published in 1637: *La géométrie, La dioptrique,* and *Les météores.* The main distinction between Bernard's book and Descartes' is that the latter contains practically no references to the author's own discoveries, which were of considerable moment. Indeed, Bernard's book is unique in the literature of philosophy of science to the extent that it draws so heavily on the author's research. In the history of science, this book is equally outstanding as an analysis of a man's own research illuminated by deep philosophical insight. Reading this book one is tempted to make the generalization that sound philosophical analysis of science can be made only by a practising scientist and the greater the scientist the better.

Claude Bernard is remembered today for four major contributions to physiology: (i) the discovery of the vasomotor nerves, (ii) the nature of the action of curare and other poisons on neuromuscular transmission, (iii) the functions of the pancreatic juice in digestion, and (iv) the elucidation of the glycogenic function of the liver. Bernard thought the last of these to be especially important because it tended to show a similarity between animals and plants, both of which can produce sugar. Today this aspect of the discovery has dwindled in importance. Of far greater significance is the new concept that arose from the investigations of the liver and expressed in these words:

FOREWORD

"The account of the liver shows very clearly that there are *internal secretions,* that is to say, secretions whose products, instead of being poured out to the exterior, are transmitted directly into the blood."

In discussing the advance of science, Bernard said: "We usually give the name of discovery to recognition of a new fact: but I think that the idea connected with the discovered fact is what really constitutes the discovery. Facts are neither great nor small in themselves." In the case of this "discovery," the "fact" of the action of the liver is of considerable importance, but the "idea connected with the observed fact"—that there are in the animal body these organs of internal secretion—has proved to be of far-reaching consequence. Bernard later included among organs of this type the thyroid and adrenals, today considered as highly typical of endocrines or organs of internal secretion. At present, curiously enough, according to J. M. D. Olmsted, the glycogenic function of the liver is "considered a special arrangement for storage and liberation of carbohydrate," rather than "as an internal secretion."

Dr. John F. Fulton has said that Bernard "left his mark on so many branches of the subject [of physiology] that it is impossible to say which of his many discoveries was the most significant." Perhaps his greatest contribution was the concept of the *milieu intérieur,* of which a complete statement appears in his last published work, and which he described as the "basis of general physiology." Bernard held that the "various ways in which living organisms are related to their cosmic environment enable us to study life in three forms;" the three forms of life are—(i) *Vie latente,* "where life is not evident," (ii) *Vie oscillante,* "where evidences of life are variable and dependent upon external environment [*milieu extérieur*]," (iii) *Vie constante,* "where life manifests itself independently of the external environment." The third of these, found in "the more highly organized animals," is "characterized by freedom and independence." Here the steady flow of life appears to be independent of "alterations in its cosmic environment or changes in its material surroundings." Enclosed "in a kind of hot-house," the organism is not affected by the "perpetual changes of external conditions, . . . but is free and independent." Bernard said:

I think I was the first to urge the belief that animals have really two environments: a *milieu extérieur* in which the organism is situated, and a *milieu intérieur* in which the tissue elements live. The living organism does not really exist in the *milieu extérieur* (the atmosphere if it breathes, salt or fresh water if that is its element) but in the liquid *milieu intérieur* formed by the circulating organic liquid which surrounds and bathes all the tissue elements; this is the lymph or plasma, the liquid part of the blood which, in the higher animals, is diffused through the tissues and forms the ensemble of the intercellular liquids and is the basis of all local nutrition and the common factor of all elementary exchanges. A complex organism should be looked upon as an assemblage of simple organisms which are the anatomical elements that live in the liquid *milieu intérieur*.[2]

Bernard concluded that the "primary condition for freedom and independence of existence" is the stability of the *milieu intérieur;* "the mechanism which allows of this is that which insures in the *milieu intérieur* the maintenance of all the conditions necessary to the life of the elements." Hence "simple organisms whose constituent parts are in direct contact with their cosmic environment" can have "no freedom or independence of existence," which is therefore "the exclusive possession of organisms which have attained a higher state of complexity or organic differentiation." Clearly, the organism must be "so perfect that it can continually compensate for and counterbalance external variations." Hence Bernard concluded that the higher animals are far from being "indifferent to their surroundings," but must be "in close and intimate relation to it." The equilibrium they maintain is "the result of compensation established as continually and exactly as if by a very sensitive balance." Bernard thus exposed the fallacy of those who held that in living organisms there exists "a free vital principle which fights against the influence of physical conditions." Quite the contrary. "All the vital mechanisms, varied as they are, have only one object: that of preserving constant the conditions of life in the *milieu intérieur*." There can be little doubt that this is Bernard's greatest biological generalization. Concerning it, L. J. Henderson has written:

[2] Quoted from the translation in Dr. John F. Fulton's *Selected Readings in the History of Physiology,* page 308.

FOREWORD

Stability may sometimes be afforded by the natural environment, as in sea water. In other cases an integument may sufficiently temper the external changes. But by far the most interesting protection is afforded, as in man and higher animals, by the circulating liquids of the organism, the blood plasma and lymph, or, as Claude Bernard called them, the *milieu intérieur*. In his opinion, which I see no reason to dispute, the existence and the constancy of the physico-chemical properties of these fluids is a necessary condition for the evolution of free and independent life. This theory of the constancy of the *milieu intérieur* was an induction from relatively few facts, but the discoveries of the last fifty years and the introduction of physico-chemical methods into physiology have proved that it is well founded.

The influence of this idea may be traced in the writings of J. S. Haldane, L. J. Henderson, Walter B. Cannon, Sir Joseph Barcroft, and others, and it has been described by J. F. Fulton as one that "will undoubtedly exert a great influence on the physiology of the future."

One of the attractive features of Claude Bernard's *Introduction to the Study of Experimental Medicine* is the frankness with which he examines the roles of chance and of error and even preliminary false conclusions in leading eventually to scientific truth. Equally valuable is his discussion of the use of hypotheses. He is said to have told Paul Bert that on entering the laboratory to perform the actual job of experimenting, he should leave his imagination in the coat room with his overcoat, but he must not forget to put it on again when he went out. His colleague Bertholet described Bernard in terms of his "sincere zeal for science, his absolute freedom from false pretension, his unsleeping spirit of curiosity, and the surety of method that he applied to his investigations."

Written in the great tradition of the French positivist and near-positivist scientific philosophy of the nineteenth century, Bernard's book still asks the scientist and the reader interested in science to re-examine fundamental concepts and the generally accepted foundations of scientific knowledge. In a day when nearly all of the literature on the philosophy of science deals exclusively with physics and mathematics, we are reminded that biology too deals with problems of basic concern to the philosopher and that philosophies of science that ignore biology must of necessity be incomplete and thereby inadequate.

* * *

Two biographies of Bernard are available in English, Michael Foster's *Claude Bernard* (London: T. Fisher Unwin, 1899) and J. M. D. Olmsted's *Claude Bernard, Physiologist* (New York and London: Harper and Brothers, 1939). An excellent account of Bernard's work in physiology is available in John F. Fulton's *Physiology* ("Clio Medica" series, vol. 5, New York: Hoeber, 1931). A selection in English translation from Bernard's *Leçons sur les phénomènes de la vie communs aux animaux et aux végétaux* (2 vols., Paris: Baillière, 1878), containing a statement concerning the *milieu intérieur* may be found in Fulton's *Selected Readings in the History of Physiology* (Springfield, Illinois: Charles C Thomas, 1930, pp. 307ff.). A critical discussion of Bernard's philosophy of science is given in Max Black's "The Definition of Scientific Method," *Science and Civilization,* ed. by Robert C. Stauffer (Madison: University of Wisconsin Press, 1949, pp. 67ff.).

Bernard had intended to produce a "Principles of Scientific Medicine" to supplement the present work. His incomplete essay was published with an introduction and notes by Léon Delhoume, *Principes de médecine expérimentale* (Paris: Presses Universitaires de France, 1947). Also of importance are Bernard's *Pensées: notes d'etachées,* ed. by Delhoume (Paris: Baillière, 1937), a manuscript entitled *Philosophie,* edited by Jacques Chevalier (Paris: Boivin, 1937-??), and the volume containing Bernard's notes on Comte's *Cours de philosophie positive* edited by E. Dhurout, *Claude Bernard: Extraits de son oeuvre* (Paris: Librairie Félix Alcan, 1939), also noteworthy for the inclusion of Henri Bergson's essay on Bernard's philosophy.

14 January 1957   I. BERNARD COHEN

INTRODUCTION

The discoverer of natural knowledge stands apart in the modern world, an obscure and slightly mysterious figure. By the abstract character of his researches his individuality is obliterated; by the rational form of his conclusions his method is concealed; and at best he can be known only through an effort of the imagination. This is perhaps inevitable. But the unfortunate effects are enhanced by convention which to-day prescribes a formal, rigorous and impersonal style in the composition of scientific literature. Thus while it is no more difficult to know Galileo and Harvey than Cervantes and Milton through their writings, or to perceive their habits and methods of work, psychological criticism will often seek in vain the personality and the behavior of the person behind the modern scientific printed page. Yet whoever fails to understand the great investigator can never know what science really is.

Such knowledge is not taught in the schools. Even more than the scientific memoir, the treatise and the lecture are formal, logical, systematic; thus truly intelligible and living only to the initiated. As much as possible science is made to resemble the world which it describes, in that all vestiges of its fallible and imaginative human origin are removed. Since the publication of Euclid's immortal textbook this has been the universal and approved usage. Little doubt should remain that it is the best. But then the burden must fall upon the student of initiating himself into mysteries which no one will explain to him.

What he lacks is understanding of the art of research and of the inevitable conditions and limitations of scientific discovery, an understanding, in short, of the behavior of the man of genius, not a rationalized discussion of scientific method. The latter may be sought in many learned works and in the teachings of academic philosophers; a good account of the former is far to seek. It is, therefore, not the least of the merits of Claude Bernard's An Introduction to the Study of Experimental Medicine that we have here an honest and successful analysis of himself at work by one of the most intelligent of modern scientists, a man of genius and a great physiologist. This work lays bare, so far as that is possible, what others have concealed.

With due regard to such analysis and logical formulation as are indispensable for intelligibility of exposition, Claude Bernard has avoided *a posteriori* rationalization as he has *a priori* dogmatism. Thus it is possible to perceive his scientific method as the habit of the man. His life is spent in putting questions to nature. These questions are the measure of his originality. He cannot tell how they arise, but the experimental idea seems to him a presentiment of the nature of things. Such ideas are, at any rate, the only fertilizing factor in research; without them scientific method is sterile, and great discoveries are those which have given rise to the most luminous ideas.

The experiment, accordingly, is always undertaken in view of a preconceived idea, but it matters not whether this idea is vague or clearly defined, for it is but the question, vague or otherwise, which he puts to nature. Now, when nature replies, he holds his peace, takes note of the answer, listens to the end and submits to the decision. In short, the experiment is always devised with the help of a working hypothesis; the resulting observation is always made without preconceived idea. Such habits are not too easily formed, for man is by nature proud and inclined to metaphysics, but the practice of experimentation will cure these faults.

Claude Bernard is at pains to point out that even so modest an abstract description of method does violence, for the sake of clearness, to the complexity of human behavior. Beyond this his method is the *art* of experimentation, an art which rests upon a perfect and habitual familiarity with the objects that he studies and with the details of his experimental procedure.

The chapters in which all this is developed are pervaded by a spirit of honesty, simplicity and modesty, the mark of a great investigator. It is not difficult while reading them to see the man at work, full of ideas, a marvelous observer, marking and taking note even of that for which he is not looking, always doubting, but serenely and without scepticism, guarding himself from his hypothesis and even from the unconfirmed observation, yet ever confident in the determinism of nature and therefore in the possibility of rational knowledge.

The subject of his investigations was physiology, in the broadest and in the most modern sense, physiology conceived as the predestined foundation of scientific medicine and as the most important part of biology. Thus his science was seen by Claude Bernard with clear but prophetic vision, for he lived almost a half century before his time. He perceived that physiology rests securely upon the physico-chemical sciences, because all that these sciences bring to light is true of organic as of inorganic phenomena. Also there is nothing but the difficulty of the task to hinder the reduction of physiological processes to physical and chemical phenomena. And yet this cannot be the last word, for physiology is more than bio-physics and bio-chemistry, biology more than applied physical science. He has himself, elsewhere, put the case as follows:

"Admitting that vital phenomena rest upon physico-chemical activities, which is the truth, the essence of the problem is not thereby cleared up; for it is no chance encounter of physico-chemical phenomena which constructs each being according to a pre-existing plan, and produces the admirable subordination and the harmonious concert of organic activity.

"There is an arrangement in the living being, a kind of regulated activity, which must never be neglected, because it is in truth the most striking characteristic of living beings. . . .

"Vital phenomena possess indeed their rigorously determined physico-chemical conditions, but, at the same time, they subordinate themselves and succeed one another in a pattern and according to a law which pre-exist; they repeat themselves with order, regularity, constancy, and they harmonize in such manner as to bring about the organization and growth of the individual, animal or plant.

"It is as if there existed a pre-established design of each organism and of each organ such that, though considered separately, each physiological process is dependent upon the general forces of nature, yet taken in relation with the other physiological processes, it reveals a special bond and seems directed by some invisible guide in the path which it follows and toward the position which it occupies.

"The simplest reflection reveals a primary quality, a *quid proprium* of the living being, in this pre-established organic harmony." *

---

* *Leçons sur les Phénomènes de la Vie Commune aux Animaux et aux Végétaux.* Paris, 1878, Vol. 1, p. 50.

I know of no other statement of the case since Aristotle's which seems to me to present so well a biologist's philosophy.

It must not be expected, however, to find in the work of Claude Bernard a *system* of biological philosophy. He sets forth his views on the philosophy and the method of science, and they are really his views, the very convictions that he carries with him into the laboratory. But they are not a clear system of philosophy, nor a rational and logical scientific method, which neither he nor anyone else can believe in as he goes about his daily work. Hence, like everybody's real beliefs, they shade off into vague, more or less inconsistent, more or less doubtful opinions. This is reality itself.

The theory of organism is more than a philosophical generalization; it is a part of the working equipment of the physiologist, fulfilling a purpose not unlike that of the second law of thermodynamics in the physical sciences. It has been more or less clearly understood and employed from the earliest times, and Claude Bernard did but perfect it. The theory of the constancy of the internal environment, a related theory, we owe almost wholly to Claude Bernard himself. There is no better illustration of his penetrating intelligence. A few scattered observations on the composition of blood sufficed to justify, in his opinion, the assertion that the constancy of the internal environment (*milieu intérieur*) is the condition of free and independent life.[1] A large part of the physiological research of the last two decades may fairly be regarded as a verification and illustration of this theory, which, as Claude Bernard perceived, serves to interpret many of the most important physiological and pathological processes. It was this theory too that led him to a clear conception of general physiology, which he regarded as the fundamental biological science.

General physiology, according to him, includes the study of the physico-chemical properties of the environment of the cell, a similar study of the cell itself, beyond this of the physico-chemical relations between cell and environment, and, generally, of the phenomena common to animals and plants. This science, of which he is the founder, was destined to remain undeveloped until long after his death. To-day, with the aid of a physical chemistry unknown to the

---

[1] This should not be thought of as absolute constancy, and it should be understood that variations in the properties of the internal environment may be both cyclical and adaptive, that is functional, but in general may not be random and functionless. Claude Bernard's principle is the first approximation which suffices until the subject has been broadly developed.

contemporaries of Claude Bernard, it is fulfilling the promise which he alone could clearly see. He never had a more luminous presentiment of the nature of things than this vision of the future foundations of biology.

No man is a true prophet otherwise than through the possession of such intimate knowledge of a subject that he is able to say, "Thus matters must develop." Such was Claude Bernard's prophecy of the future of his own science. His understanding of physiology had become so perfect that the future could not be wholly doubtful. He knew where the path must lead, and it is this that makes his book so amazingly modern. In other respects he is only a highly intelligent man of the third quarter of the nineteenth century. Accordingly his treatment of some subjects, such as mathematics and physics, is a little old-fashioned, especially on the logical side. In general such defects are not only slight, but also unimportant from the medical standpoint. But his discussion of statistics could hardly be written to-day. There are indeed those, though few in number, who will agree with his criticisms. But, when he wrote, the influence of Galton had not been exerted and nobody realized that statistics afford a method, at once powerful, elegant and exact, of describing a class of objects as a class.

Physiology, as defined and understood in this book, with general physiology as its foundation, is the essential medical science. Medicine has passed through the empirical, the systematic, the nosological and the morphological stages and has entered upon the experimental stage. Thus it has finally become physiological, for physiology is the larger part of experimental medicine. Such is the principal thesis of the present work, which ought not to be obscured by the consideration of incidental topics, no matter how intrinsically important they may be.

This opinion, to be sure, does not yet meet with universal approval, and yet I believe that it has been at length fully confirmed by the experience of the twentieth century. Nevertheless, the confirmation was long delayed by the emergence of the bacteriological stage in the evolution of medicine. Unforeseen by Claude Bernard, this was the result of the discoveries of his contemporary, Pasteur.

To-day, looking backward, we see how it was that bacteriological researches for a long time took the first place which Claude Bernard believed to be already assured to those of his own science. When

Pasteur began the study of micro-organisms a great gap existed in our knowledge of the organic cycle and of natural history. His work and that of his successors filled this gap, completed our present theory of the cycle of life and established the natural history of infectious diseases, of fermentations and of the soil. This was perhaps the most rapid advance of descriptive knowledge in the history of science. For the moment the researches of physiologists were overshadowed and the work of the young men diverted into the new fields. In time bacteriology grew into a fully developed science, perfected its methods, exploited its domain, and then, the most pressing work well done, resigned its leadership of the medical sciences.

Meanwhile a profound influence was exerted on what Professor Whitehead has called the intellectual climate. Claude Bernard's outlook may be described as biological and philosophical, and such a point of view seems necessary for the understanding of the deeper problems of medicine. Pasteur, however, always retained the chemist's outlook, and in him the will was more important than the reflective intellect. His successors have taken a position hardly more biological and, probably of necessity, have had little interest in rational theory. Such a climate is unfavorable to the growth of experimental medicine and especially of general physiology, for both are biological and rational.

This had been vaguely understood as early as the times of Galileo, of Borelli, and of Malpighi, when the minds of men were still fresh and not yet enslaved by specialism. But even Claude Bernard, because he still lacked the aid of modern physical chemistry, hardly appreciated the possibilities, very limited but very important, of the applications of the fully developed method of rational physical science, when guided and duly restrained by the judgment of a true physiologist, in the study of the ultimate phenomena of life.

In default of the physico-chemical foundations, during a period when bacteriology was the dominant influence in medical science, and next to it, perhaps, the highly specialized science of organic chemistry, when the prevailing activity was somewhat unintellectual, physiology continued in the old paths. Not until after the turn of the century did the movement which Claude Bernard had foreseen make itself felt. To-day it is well established and should be generally recognized. The result has already been a remarkable increase of experimental investigation and of rational theorizing in the clinic.

For the first time mathematics, physics, chemistry and physical chemistry, as aids to physiology, have passed into the hospitals. I believe that, for the reasons which Claude Bernard has explained, this will long remain the way of medical progress and that we have now definitely entered upon the epoch of experimental medicine.

All progress entails evils and few experimenters can understand as Claude Bernard did the phenomena of life and the philosophy of the organism. For these reasons, and for others not so good, the growth of experimental medicine gives rise to criticism, as it did a half century ago. Experienced physicians, practised in the art of medicine and rightly believing that medicine is still and must always be an art, but also uncomfortable and suspicious through ignorance of the new development, are not lacking to unite with this opposition. So far as grounds for complaint exist they are due to the absence of that high intelligence and skill of the experimenter which are necessary to understand and to solve the complex problems of physiology. Here one can only plead the palliating circumstance that all human endeavor suffers from the same weaknesses. On the other hand, prevailing criticism of scientific medicine itself, no less than the earlier criticism of the nineteenth century, finds conclusive answer in this book.

Medicine is but a part of human biology and the study of human inheritance, constitution, intelligence and behavior, of adaptation to new conditions of life, and of a host of other subjects, far transcends the boundaries of medicine. But everywhere throughout this vast field physiology has the same importance as in the narrower field of medicine. Thus the Introduction may serve as a guide not only for those who are beginning the study of medicine, but for many others as well.

The sciences are not equal, nor do they preserve their rank unchanged as civilization moves on. During nearly a quarter of a millennium mechanics led all the others in intellectual interest and in influence upon European civilization. It will seem to many not too bold a prophecy, for the reasons that Claude Bernard has set forth, to look forward to a century in which physiology shall take a similar place. I venture to believe that that position will be reached when the experimental method has made possible a rational science of organism.

The physiological researches of Claude Bernard have immortalized his name, but the present work and his other general writings have hardly attracted the attention which they deserve or exerted the influence of which they are capable. This is probably due both to the conflicting influence of bacteriology, organic chemistry and other sciences and, not less, to his own clearness of vision. That which he saw as the future of physiology remained for many decades hidden from others and so his writings were only half understood. Even general physiology is still hardly aware of the program which he set forth and which it has been unwittingly carrying out. There is, however, one well known instance of his influence exerted farther afield. As the idea of Balzac's *Comédie Humaine* was suggested by the biology of the early nineteenth century, so the naturalism of Zola was suggested by the works of Claude Bernard. Perhaps the result will not be thought worthy of the cause. Yet the instance is significant of the wide bearing of an interpretation of life which may be seen to be peculiarly well suited to the present conditions of the political and social as well as of the natural sciences.

Among great men, Claude Bernard should be counted fortunate in that he has not become a mythical figure. Unlike Pasteur, whose discoveries are hardly more remarkable, though their immediate influence has been immeasurably greater, and whose horizon was incontestably less broad, he remains a plain man, highly distinguished no doubt, but not obscured by the growth of a legend.

It is possible not only to see him at work, but even to discover his purposes and his feelings. The desire to relieve suffering and a sense of duty are clearly apparent, and one may read between the lines the enduring satisfaction that he felt in the society of younger men who owed to him more than they could ever repay. But weightier still are the contentment which comes from work well done, the sense of the value of science for its own sake, insatiable curiosity and, above all, the pleasure of masterly performance and of the chase. These are the effective forces which move the scientist. The first condition for the progress of science is to bring them into play.

L. J. HENDERSON.

October 11, 1926

CLAUDE BERNARD

CLAUDE BERNARD, born July 12, 1813, at Saint Julien near Villefranche,[1] came to Paris in 1832, with almost no paraphernalia except a tragedy which had never been acted, and a farce-comedy which had had some success at a small theatre in Lyons. He showed these first attempts to Saint-Marc Girardin who was temporarily taking Guizot's place at the Sorbonne. Girardin advised him to learn a profession to live by, and to write poetry in his spare time: certainly he had no idea that standing before him was a future colleague in the French Academy. Young Claude Bernard followed this sensible advice and entered the school of medicine.

Though he received his appointment as hospital interne in 1839, he was anything but a brilliant pupil. His comrades did not suspect what lay hidden behind the huge forehead of this silent student who paid so little attention to his professors' teaching, that they easily condemned his meditative calm as mere laziness. Survivors remember and often describe that revelation, his publications on gastric sugar, the chorda tympani, the pneumogastric nerve and the spinal nerve, which suddenly revealed to the scientific world a sagacious and ingenious experimenter of rare operative skill.

Magendie's teaching brought this revolution about. As soon as Claude Bernard set foot in the laboratory of the Collège de France, his path was marked out. The celebrated physiologist's daring though somewhat disorderly experimentation, his pitiless criticism, the scepticism that included even his own discoveries, made a deep and, so to speak, creative impression on the young man's mind. But the pupil was so much more powerful than his master that he took from his teaching only its virtues of independence, and succeeded in keeping doubt within scientific bounds. To deep disdain for plausible explanations in which alluring chimeras are concealed, he easily added respect for the facts gathered in tradition, sincere belief when face to face with the unexpected which is often pregnant with discovery, respect for searching hypotheses and coördinating theories,

---

[1] Department of the Rhone.

without ever attributing to them independent authority or power. Finally what distinguished him especially from Magendie, and gave him his wholly individual character, was love of certainty,—that deep feeling for law, that immovable confidence that,—if the conditions in which vital phenomena come to pass are infinitely many, complex and hard to grasp, assemble and master experimentally,— they are none the less surely and fixedly linked to phenomena without any possibility of a *quid divinum* being invoked to explain the seemingly spontaneous irregularities which they present.

This is the main point where Claude Bernard showed his superiority, from the first moments of his scientific life. The pupil of Magendie, the sceptic, introduced determinism into the realm of physiology. Thanks to him, the scientific method, respect for whose laws leads to certainty in the sciences of dead matter, assumed equal authority in the sciences of living beings. Sciences are not of two kinds, the first proud and confident, the rest timid and hesitant; the first sure of commanding and of being obeyed in experiments, the rest always in fear of an influence unknown in essence, force and goal.

It required no small effort to banish this menacing unknown from the field of physiology. The most celebrated of French physiologists, Bichat, had given it asylum, and everyone after him had thought it necessary to reckon with this capricious force, with these vital functions, whose rôle was to resist the universal laws of matter, which thus made all acts performed by living beings a series of miracles. Of course Magendie was not the man to let himself be frightened by this ghost; but he systematically and artificially simplified facts so that he only partly mastered them; or else the multiplicity of conditions governing vital phenomena took away all his theoretical confidence in the result. Now, without results there can be no science. I must repeat that Claude Bernard, therefore, proved himself, almost from the outset, superior to both Magendie and Bichat, since he felt not only the endless multiplicity of unknown data in physiology, but also their subordination to the general laws of matter and their obedience to the experimental method.

Physiology could therefore extend its roots into the solid earth where its older sisters, physics and chemistry, are settled. The complexity of the problems made it essential, however, to set forth the rules of the experimental method in special formulæ with a view to

the intellectual and manual methods which are especially adapted
to physiology. Through the whole first phase of his scientific life,
Claude Bernard was haunted by this task. But the fascination of
his laboratory and the hunt for discoveries so completely absorbed
his time that he could demonstrate the experimental method only
as Diogenes demonstrated motion.

And never was hunt for discoveries more fruitful. In twenty
years, Claude Bernard found more dominating facts, not only than
the few French physiologists working beside him, but than all the
physiologists in the world. The activity of different glands and
particularly of the pancreas, animal glycogenesis, experimental production of diabetes, the existence of the vasomotor nerves and the
theory of animal heat, the influence of poisons, studied in themselves
and as a means of analyzing physiological phenomena, the endless
number of fresh facts, keen deductions and ingenious and suggestive
insights, contained not only in his special memoirs but in the four-
teen volumes, from his *Lessons in Experimental Physiology Applied
to Medicine* (1885-1886) to his *Lessons on Diabetes and Animal
Glycogenesis* (1877), in which he collected each year the results of
his investigations and a summary of his courses,—these things gave
him the position of a master unquestioningly accepted in France and
abroad.

In official life he also attained the highest rank. In 1854, a
chair of general physiology was founded for him at the Sorbonne,
which in 1868 he surrendered, with beautiful magnanimity and
grace, to his pupil, Paul Bert. In 1858 he took Magendie's place
in the chair of medicine at the Collège de France. Member of
the Academy of Sciences in 1854, he was called in 1868 to take Flou-
rens' seat in the French Academy. Finally in 1869, by special de-
cree, he entered the Senate; and he was almost the only member of
that assembly whom no one ever thought of reproaching for a nomi-
nation which to him was so strange a surprise.

A few years before these unexpected literary and political honors
thus sought him out in his laboratory, a serious event occurred in his
life. A long and severe disease, during which he and his friends
despaired of a favorable outcome, condemned him to physical inac-
tivity. He had to leave his laboratory, and Paris too; he had to
ask of his birthplace, once more and not in vain, the gift of life and
health. Long months of isolation and rest gave back all his liberty

of mind. For the first time, he had leisure for meditation and for
setting in order, on paper, the results of his solitary reflections. A
short preface, which was already in proof and which was to have
preceded a sort of treatise on operative physiology which remains
unfinished, grew, by successive additions, to the size of a pamphlet,
then of a book which saw the light in 1865.

The *Introduction to the Study of Experimental Medicine* struck
cultivated minds with admiration and astonishment. Here physiologists were happy to find, reduced to precise formulæ, set in order
with marvellous art, and lighted by examples which themselves were
like so many intellectual experiments,—here they were happy to find
the rules of the experimental method, watching, seizing and, in spite
of its struggles, mastering that organic Proteus of the deceitful
metamorphoses. Men less taken up with professional difficulties
were struck by the magnitude of the problems studied, by the clarity
of exposition, the ease and good faith with which they were either
solved or proved insoluble. Even the style attracted great attention;
its original flavor took even the French Academy's fancy: "You have
created a style," said the severe Monsieur Patin, in his speech of
welcome. And it was true. But how surprised the venerable critic
would have been if he had read the earlier books in which Claude
Bernard contented himself with enumerating his laboratory impres-
sions in a narrative that is often scarcely well ordered. The eminent
but naïve master was never haunted by care for stage effect; his style,
whether spoken or written, is the equivalent of his ideas. In epi-
sodical narrative, he is often dragging and confused; but when a
hard problem presents itself, when his thought is forced to fall back
as if to conquer an obstacle or make a bound, then he concentrates,
purifies and accentuates himself in definite formulæ and often in
verbal imagery.

As he was in his books, so was Claude Bernard in his courses and
his conversation. His was not a docile thought, speaking every
language and playing every rôle; and he never disciplined it to any
conventions of profession or of social custom. If it escaped him, he
followed it without rebellion, leaving his speech drooping, his lecture
in confusion, while he listened to what it softly said to him. But
if it grew interested in the immediate subject, then the professor or
conversationalist, a moment ago so difficult and diffuse, awoke living,
inventive, clear, eloquent, with surprising and sudden changes, and

always with both characteristics of true genius, ease and good faith.

And no one possessed them in higher degree. That ease in raising himself to high summits, in acting in the midst of trying difficulties, especially struck readers of his admirable articles in the *Revue des Deux Mondes.* What the poet said of the goddess could be said of him: *incessu patuit.* After reading these articles, an eminent man said to me one day: "He does not make me merely think I understand, as you all do; he makes me really understand." And in fact he did understand. Claude Bernard carried this ease from his physiological method into the philosophic realm. No one ever made discoveries more simply, more naïvely. To that first phase of hunting ideas, which consists, as Helvetius said, in seeing and starting the quarry, he brought a sureness of vision, an astounding penetration. Most scientific searchers are a kind of somnambulist who see only what they are looking for and what is on the track of their ideas; their eyes are fixed on a point; and they fail to perceive not only what happens aside from that point, but even what appears there unforeseen. In one of his pupil's phrases, Claude Bernard seemed to have eyes all around his head. In the course of an experiment, students were stupefied when they saw him point out quite evident phenomena which no one but himself had seen. He discovered as others breathed.

With ease, good faith. That was his ruling characteristic. He never swerved from the deep sincerity of a man of science who must seek truth for its own sake and for the truths which follow from it, without concerning himself with the distant or indirect conclusions which lawyer-like men, with a cause to defend, try to draw from it. No one was ever more passive in deduction, or described deductions with more candid sincerity. This is why supporters of different theses could use, and still can use, his writings, turn and turn about. When he expounds the cerebral determinism of intellectual activity, the materialists count him as their own; when he declares that thought and the brain are in the same relation as time and a clock, the spiritualists try to enlist him. In reality, he is just a physiologist handing over fresh facts to rejuvenate the speculator's endless wrangling.

In the narrower realm of physiology and medicine his admirable good faith explains the seeming contradiction between his scientific faith and his practical incredulity. He always had this double feeling in the highest degree,—that if medicine is to be sure of itself, it must have physiology as its necessary base, and that our present-day physiology is still far from supplying us with any practical certainty. He felt the full importance of his own discoveries as foundations for the medical edifice, but he did not share the illusions of those whose eagerness to transfer them to the realm of clinical or therapeutic applications often made him smile. The feeling for distances, which would have discouraged less valiant men, moved him not at all. For strength and perseverance, he did not need the intoxication of illusions. So he, who taught that medicine is or should be a science, showed himself thoroughly sceptical about physicians; and when he talked of them, the shade of Sganarelle [2] always seemed to pass before him.

The *Introduction to the Study of Experimental Medicine* marks a fresh phase in the life of Claude Bernard. From this period date the philosophic writings which opened the French Academy's doors to him. Of this period, too, are the books [3] in which grouping facts takes precedence over noting details, and in which he returns to his earlier discoveries and strives to bring his subsidiary work to the precision and perfection which present-day science permits.

This does not mean that he turned completely away from those regions of the unknown where he had formerly reaped such rich harvests. His latest work on the fundamental identity of the properties of tissues and of elementary functions in the animal and vegetable kingdoms, on anesthesia of the lower vegetables by chloroform or ether, and on the general action of toxic substances, shows that the creative spirit was still alive in him.

Fresh discoveries were this year to have furnished another proof of his fertile activity. He confided this in part to his friends and pupils; and from the few words which escaped him, we may apparently conclude that the investigations which he carried out during his last vacation were to throw unexpected light on the theory of fermentation. This important work, of which he said, only four days ago, "What a pity; it would have been good to finish it!" is lost to science.

[2] In *Le Médecin malgré lui* by Molière.
[3] *Recherches sur les propriétés des tissus vivants. Leçons de pathologie expérimentale*, etc.

On December 31 he was stricken in the laboratory of the Collège de France; shivering and fever soon came on, and special phenomena indicating inflammation of the kidneys. Nothing could stop the advance of a disease whose every progress he followed. Without any illusions about the fatal catastrophe, he observed with calm eyes, and with a smile denied his scientific family's pious lies. He was one of those whose gaze is undismayed by the unknown.

Personal feeling must be silent in this immense mourning of science, and yet, the loss of a great man is not all that moistens the eyes of those about his coffin: such kindliness, such simplicity of soul, such naïve generosity were united in his genius. One's hand trembles in trying to sketch a few traits of this great and noble character.

Nothing in his pure and harmonious life was turned aside from its chief aim. Enamored of literature, art and philosophy, Claude Bernard as a physiologist lost nothing by these noble passions; on the contrary, they all helped in developing the science with which he identified himself, and of which he is the highest and most complete embodiment. He was a physiologist such as no man had been before him. "Claude Bernard," said a foreign scientist, "is not merely a physiologist, he is physiology."

His very death seems to mark a new era in science. For the first time in our country, a man of science will receive those public honors hitherto reserved for political and military celebrities. The cabinet honored itself yesterday in asking parliament, which unanimously agreed to celebrate at state expense the solemn obsequies of the master who is no more. And one phrase of Gambetta, speaking in the name of the Budget Commission, sums up all that we have said: "The light, which has just been extinguished, cannot be replaced."

PAUL BERT.

Paris, February 12, 1878.

CONTENTS

AN INTRODUCTION TO THE STUDY OF EXPERIMENTAL MEDICINE

To CONSERVE HEALTH AND TO CURE DISEASE: Medicine is still pursuing a scientific solution of this problem, which has confronted it from the first.[1] The present state of medical practice suggests that a solution is still far to seek. During its advance through the centuries, however, medicine has always been driven into action and from numberless ventures in the realm of empiricism has gained useful information. Though furrowed and overturned by all manner of systems so evanescent that, one by one, they have disappeared, it has none the less carried on research, acquired ideas and piled up precious materials which in due time will find their place and meaning in scientific medicine. To-day, thanks to the great development and powerful support of the physico-chemical sciences, study of the phenomena of life, both normal and pathological, has made progress which continues with surprising rapidity.

It is therefore clear to all unprejudiced minds that medicine is turning toward its permanent scientific path. By the very nature of its evolutionary advance, it is little by little abandoning the region of systems, to assume a more and more analytic form, and thus gradually to join in the method of investigation common to the experimental sciences.

In order to embrace the medical problem as a whole, experimental medicine must include three basic parts: physiology, pathology and therapeutics. Knowledge of causes of the phenomena of life in the normal state, i.e., PHYSIOLOGY, will teach us to maintain normal conditions of life and to conserve HEALTH. Knowledge of diseases and of their determining causes, i.e., PATHOLOGY, will lead us, on the one hand, to prevent the development of morbid conditions,

---

[1] See *Cours de pathologie expérimentale* (*Medical Times*, 1859-1860).—*Leçon d'ouverture du cours de médecine du Collège de France: Sur la médecine expérimentale* (*Gazette médicale*. Paris, April 15, 1864;—*Revue des cours scientifiques*. Paris, Dec. 31, 1864.)

and, on the other, to fight their results with medical agents, i.e., to cure the diseases.

In the empirical period of medicine, which must doubtless still be greatly prolonged, physiology and therapeutics could advance separately; for as neither of them was well established, they were not called upon mutually to support each other in medical practice. But this cannot be so when medicine becomes scientific: it must then be founded on physiology. Since science can be established only by the comparative method, knowledge of pathological or abnormal conditions cannot be gained without previous knowledge of normal states, just as the therapeutic action of abnormal agents, or medicines, on the organism cannot be scientifically understood without first studying the physiological action of the normal agents which maintain the phenomena of life.

But scientific medicine, like the other sciences, can be established only by experimental means, i.e., by direct and rigorous application of reasoning to the facts furnished us by observation and experiment. Considered in itself, the experimental method is nothing but reasoning by whose help we methodically submit our ideas to experience,—the experience of facts.

Reasoning is always the same, whether in the sciences that study living beings or in those concerned with inorganic bodies. But each kind of science presents different phenomena and complexities and difficulties of investigation peculiarly its own. As we shall later see, this makes the principles of experimentation incomparably harder to apply to medicine and the phenomena of living bodies than to physics and the phenomena of inorganic bodies.

Reasoning will always be correct when applied to accurate notions and precise facts; but it can lead only to error when the notions or facts on which it rests were originally tainted with error or inaccuracy. That is why experimentation, or the art of securing rigorous and well-defined experiments, is the practical basis and, in a way, the executive branch of the experimental method as applied to medicine. If we mean to build up the biological sciences, and to study fruitfully the complex phenomena which occur in living beings, whether in the physiological or the pathological state, we must first of all lay down principles of experimentation, and then apply them to physiology, pathology and therapeutics. Experimentation is undeniably harder in medicine than in any other science; but for that very reason, it was never so necessary, and indeed so indispensable. The more complex the science, the more essential is it, in fact, to establish a good experimental standard, so as to secure comparable facts, free from sources of error. Nothing, I believe, is to-day so important to the progress of medicine.

To be worthy of the name, an experimenter must be at once theorist and practitioner. While he must completely master the art of establishing experimental facts, which are the materials of science, he must also clearly understand the scientific principles which guide his reasoning through the varied experimental study of natural phenomena. We cannot separate these two things: head and hand. An able hand, without a head to direct it, is a blind tool; the head is powerless without its executive hand.

The principles of experimental medicine will be explained in this work from the triple point of view of physiology, pathology and medicine. But before going into general considerations and special descriptions of the operative procedure proper to each of these divisions, I deem it useful to give a few explanations in this introduction in relation to the theoretic and philosophic side of the method which this book, after all, treats merely on its practical side.

The ideas which we shall here set forth are certainly by no means new; the experimental method and experimentation were long ago introduced into the physico-chemical sciences, which owe them all their brilliancy. At different periods, eminent men have treated questions of method in the sciences; and in our own day Monsieur Chevreul, in all his works, is explaining very important ideas on the philosophy of experimental science. We shall therefore make no claim to philosophy. Our single aim is, and has always been, to help make the well-known principles of the experimental method pervade medical science. That is why we shall here recapitulate these principles, specially pointing out the precautions to be taken in their application, because of the very special complexity of the phenomena of life. We shall consider these difficulties, first in the use of experimental reasoning, and then in the practice of experimentation.

PART ONE

EXPERIMENTAL REASONING

CHAPTER I

OBSERVATION AND EXPERIMENT

ONLY within very narrow boundaries can man observe the phenomena which surround him; most of them naturally escape his senses, and mere observation is not enough. To extend his knowledge, he has had to increase the power of his organs by means of special appliances; at the same time he has equipped himself with various instruments enabling him to penetrate inside of bodies, to dissociate them and to study their hidden parts. A necessary order may thus be established among the different processes of investigation or research, whether simple or complex: the first apply to those objects easiest to examine, for which our senses suffice; the second bring within our observation, by various means, objects and phenomena which would otherwise remain unknown to us forever, because in their natural state they are beyond our range. Investigation, now simple, again equipped and perfected, is therefore destined to make us discover and note the more or less hidden phenomena which surround us.

But man does not limit himself to seeing; he thinks and insists on learning the meaning of the phenomena whose existence has been revealed to him by observation. So he reasons, compares facts, puts questions to them, and by the answers which he extracts, tests one by another. This sort of control, by means of reasoning and facts, is what constitutes experiment, properly speaking; and it is the only process that we have for teaching ourselves about the nature of things outside us.

In the philosophic sense, observation shows, and experiment teaches. This first distinction will serve as our starting point in examining the different definitions of observation and experiment devised by philosophers and physicians.

I. VARIOUS DEFINITIONS OF OBSERVATION AND EXPERIMENT

Men sometimes seem to confuse experiment with observation. Bacon appears to combine them when he says: "Observation and experiment for gathering material, induction and deduction for elaborating it: these are our only good intellectual tools."

Physicians and physiologists, like most men of science, distinguish observation from experiment, but do not entirely agree in defining the two terms.

Zimmermann [1] expresses himself as follows: "An experiment differs from an observation in this, that knowledge gained through observation seems to appear of itself, while that which an experiment brings us is the fruit of an effort that we make, with the object of knowing whether something exists or does not exist."

This definition embodies a rather generally accepted opinion. According to this definition, observation would be noting objects or phenomena, as nature usually presents them, while experiment would be noting phenomena created or defined by the experimenter. We should set up a sort of contrast, in this way, between observers and experimenters: the first being passive in the appearance of phenomena; the second, on the other hand, taking a direct and active part in producing them. Cuvier expressed the same thought in saying: "The observer listens to nature; the experimenter questions and forces her to unveil herself."

At first sight, and considering things in a general way, this distinction between the experimenter's activity and the observer's passivity seems plain and easy to establish. But as soon as we come down to experimental practice we find that, in many instances, the separation is very hard to make, and that it sometimes even involves obscurity. This comes, it seems to me, from confusing the art of investigation, which seeks and establishes facts, with the art of reasoning, which works them up logically in the search for truth. Now in investigation there may be activity, at once of the mind and of the senses, whether in making observations or in making experiments.

Indeed, if we chose to admit that observation is characterized by this alone, that men of science note phenomena which nature produces spontaneously and without interference by them, still we could

---

[1] Zimmermann, *Traité sur l'expérience en médecine.* Paris, 1774.   Vol. I, p. 45.

not conclude that the mind, like the hand, always remains inactive in observation; and we should be led to distinguish under this head two kinds of observations, some passive, others active. I assume, for instance, what often occurs,—that some endemic disease appears in a region and presents itself to a physician's observation. Here is a spontaneous or passive observation which the physician makes by chance and without being led to it by any preconceived idea. But after observing the first case, if the physician has an idea that the appearance of this disease may well be related to certain special meteorological or hygienic circumstances, he takes a journey to other regions where the same disease prevails, to see whether it develops under the same conditions. This second observation, made in view of a preconceived idea of the nature and cause of the disease, is what we must obviously call an induced or active observation. I should say as much of an astronomer who, in watching the sky, discovers a planet passing, by chance, before his telescope; in this case he makes a fortuitous or passive observation, i.e., without a preconceived idea. But, if the astronomer, after noting the aberrations of a planet, goes on to make observations, to seek a reason for them, then I should say that he makes active observations, i.e., observations produced by a preconceived idea of the cause of the aberration. We might multiply instances of this kind *ad infinitum,* to prove that, in noting natural phenomena that present themselves, the mind is now passive, now active,—which means, in other words, that observations are made, now without a preconceived idea and by chance, and again with a preconceived idea, i.e., with intention to verify the accuracy of a mental conception.

On the other hand, if we concede, as we said above, that experiment is characterized by this alone, that men of science note phenomena which they have produced artificially and which would not naturally have presented themselves, even then we could not find that the experimenter's hand always actively interfered to bring about the appearance of these phenomena. In certain cases indeed we have seen accidents where nature acted for him; and here again, from the point of view of manual intervention, we shall be forced to distinguish between active experiments and passive experiments. Let me assume that a physiologist wishes to study digestion and to learn what happens in a living animal's stomach; he will divide the walls of the abdomen and stomach according to known operative rules and will establish what is called a gastric fistula. The physiologist will certainly think that he has made an experiment, because he has interfered actively to make phenomena appear which did not present themselves naturally to his eyes. But now, let me ask, did Dr. W. Beaumont make an experiment when he came across that young Canadian hunter who had received a point-blank gun-shot in the left hypochondria, and who had a wide fistula of the stomach in the scar, through which one could look inside that organ? Dr. Beaumont took this man into his service and was able to study the phenomena of gastric digestion *de visu* for several years, as he shows in the interesting journal which he has given us on this subject.[2] In the first case, the physiologist acted on the preconceived idea of studying digestive phenomena and made an active experiment. In the second case, an accident produced a fistula of the stomach, and it presented itself fortuitously to Dr. Beaumont. According to our definition, he made a passive experiment. These examples therefore prove that, in verifying the phenomena called experiments, the experimenter's manual activity does not always come in, since it happens that the phenomena, as we have seen, may present themselves as fortuitous or passive observations.

But certain physiologists and physicians characterize observation and experiment somewhat differently. For them, observation consists in noting everything normal and regular. It matters little whether the investigator has produced the appearance of the phenomena himself or by another's hands or by accident; he considers them without disturbing them in their natural state and so makes an observation. Thus, according to these authors, observations were made in both examples of gastric fistula cited above, because in both cases we had under our eyes digestive phenomena in their natural state. The fistula served only for seeing better and making observations under the most favorable conditions.

Experiment, according to the same physiologists, implies, on the contrary, the idea of a variation or disturbance that an investigator brings into the conditions of natural phenomena. This definition corresponds, in fact, to a large group of experiments made in physiology, which might be called experiments by destruction. This form of experimenting, which goes back to Galen, is the simplest; it

by some scientist

[2] W. Beaumont, *Experiments and Observations on the Gastric Juice and on Physiological Digestion.* Boston, 1834.

should suggest itself to the minds of anatomists wishing to learn, in the living subject, the use of parts that they have isolated by dissection in the cadaver. To do this, we suppress an organ in the living subject, by a section or ablation; and from the disturbance produced in the whole organism or in a special function, we deduce the function of the missing organ. This essentially analytic, experimental method is put in practice every day in physiology. For instance, anatomy had taught us that two principal nerves diverge in the face: the facial (seventh cranial) and the trigeminal (fifth cranial); to learn their functions, they were cut, one at a time. The result showed that section of the facial nerve brings about loss of movement, and section of the trigeminal, loss of sensation, from which it was concluded that the facial is the motor nerve of the face, and the trigeminal the sensory nerve.

We said that, in studying digestion by means of a fistula, we merely make an observation, according to the definition which we are examining. But after we have established the fistula, if we go on to cut the nerves of the stomach, in order to see the changes which result in the digestive function, then, according to the same way of thinking, we make an experiment, because we seek to learn the function of a part from the disturbance which its suppression involves. And this may be summed up by saying that in experimentation we make judgments by comparing two facts, one normal, the other abnormal.

This definition of experiment necessarily assumes that experimenters must be able to touch the body on which they wish to act, whether by destroying it or by altering it, so as to learn the part which it plays in the phenomena of nature. As we shall later see, it is on this very possibility of acting, or not acting, on a body that the distinction will exclusively rest between sciences called sciences of observation and sciences called experimental.

But if the definition of experiment which we have just given differs from the definition examined in the first place in that it admits that we make an experiment only when we can vary or can dissociate phenomena by a kind of analysis, still it resembles the first in that it also always assumes an intentional activity on the experimenter's part, in producing a disturbance of the phenomena. Now it will be easy to show that the operator's intentional action can often be replaced by an accident. Here too, as in the first definition, we might distinguish between disturbances occurring intentionally and disturbances occurring spontaneously or unintentionally. Indeed taking again the example in which a physiologist cuts the facial nerve to learn its function, I assume that a ball, a sabre cut or a splinter of stone, has cut or destroyed the facial nerve; there will result fortuitously a paralysis of movement, i.è., a disturbance, exactly the same as that which the physiologist caused intentionally.

It is the same in the case of numberless pathological lesions which are real experiments, by which physicians and physiologists profit, without any purpose on their part to produce the lesions, which result from disease. I emphasize this idea now, because it will be useful to us later, to prove that medicine includes real experiments which are spontaneous, and not produced by physicians.[3]

I will make one more remark by way of conclusion. If indeed we characterize experiment by a variation or disturbance brought into a phenomenon, it is only in so far as we imply that the disturbance must be compared with the normal state. As experiments indeed are only judgments, they necessarily require comparison between two things; and the intentional or active element in an experiment is really the comparison which the mind intends to make. Now, whether the alteration is produced by accident or otherwise, the experimenter's mind compares none the less. It is therefore unnecessary to regard as a disturbance one of the facts to be compared, especially as there is nothing disturbed or abnormal in nature; everything happens according to laws which are absolute, i.e., always normal and determined. Effects vary with the conditions which bring them to pass, but laws do not vary. Physiological and pathological states are ruled by the same forces; they differ only because of the special conditions under which the vital laws manifest themselves.

II. Gaining Experience and Relying on Observation Is Different from Making Experiments and Making Observations

The general objection which I make to the preceding definitions is that they give words too narrow a meaning, by taking account of only the art of investigation, instead of considering observation and experiment at the same time as the two opposite extremes of experi-

[3] Lallemand, *Propositions de pathologie tendant à éclairer plusieurs points de physiologie.* Thesis. Paris, 1818. 2nd edition, 1824.

mental reasoning. So we find these definitions lacking in clearness and generality. To give the definition its full usefulness and value, therefore, I think that we must distinguish what pertains to the method of investigation, used to gather facts, from the characteristics of the intellectual method, which utilizes facts and makes them at once the support and the criterion of the experimental method.

In French the word *expérience* in the singular means, in general and in the abstract, the knowledge gained in the practice of life. When we apply to a physician the word experience in the singular, it means the information which he has gained in the practice of medicine. It is the same with the other professions; and it is in this sense that we say that a man has gained experience, or that he has experience. Subsequently the word *expérience* (experiment) in the concrete was extended to cover the facts which give us experimental information about things.

The word observation in the singular, in its general and abstract use, means noting a fact accurately with the help of appropriate studies and means of investigation. In the concrete the word observation has been extended to cover the facts noted; and it is in this sense that we speak of medical observations, astronomical observations, etc.

Speaking concretely, when we say "making experiments or making observations," we mean that we devote ourselves to investigation and to research, that we make attempts and trials in order to gain facts from which the mind, through reasoning, may draw knowledge or instruction.

Speaking in the abstract, when we say "relying on observation and gaining experience," we mean that observation is the mind's support in reasoning, and experience the mind's support in deciding, or still better, the fruit of exact reasoning applied to the interpretation of facts. It follows from this that we can gain experience without making experiments, solely by reasoning appropriately about well-established facts, just as we can make experiments and observations without gaining experience, if we limit ourselves to noting facts.

Observation, then, is what shows facts; experiment is what teaches about facts and gives experience in relation to anything. But as this teaching can come through comparison and judgment only, i.e., by sequence of reasoning, it follows that man alone is capable of gaining experience and perfecting himself by it.

"Experience," says Goethe, "disciplines man every day." But this is because man reasons accurately and experimentally about what he observes; otherwise he could not correct himself. The insane, who have lost their reason, no longer learn from experience; they no longer reason experimentally. Experience, then, is the privilege of reason. "Only man may verify his thoughts and set them in order; only man may correct, rectify, improve, perfect and so make himself every day more skilful, wise and fortunate. Finally for man alone does the art exist, that supreme art of which the most vaunted arts are mere tools and raw material: the art of reason, reasoning." [4]

In experimental medicine, we shall use the word experience in the same general sense in which it is still everywhere used. Men of science learn every day from experience; by experience they constantly correct their scientific ideas, their theories; rectify them, bring them into harmony with more and more facts, and so come nearer and nearer to the truth.

We can learn,—i.e., gain experience of our surroundings,—in two ways, empirically and experimentally. First there is a sort of teaching or unconscious and empirical experience, which we get from dealing with separate objects. But the knowledge which we gain in this way is also accompanied necessarily by vague experimental reasoning which we carry on quite unawares, and in consequence of which we bring together facts to make a judgment about them. Experience, then, may be gained by empirical and unconscious reasoning; but the obscure and spontaneous movement of the mind has been raised by men of science into a clear and reasoned method, which therefore proceeds consciously and more swiftly toward a definite goal. Such is the experimental method in the sciences by which experience is always gained by virtue of precise reasoning based on an idea born of observation and controlled by experiment. In all experimental knowledge, indeed, there are three phases: an observation made, a comparison established and a judgment rendered. By the experimental method, we simply make a judgment on the facts around us, by help of a criterion which is itself just another fact so arranged as to control the judgment and to afford experience. Taken in this general sense, experience is the one source of human knowledge. The mind in itself has only the feeling

[4] Laromiguière, *Discours sur l'identité: Œuvres.* Vol. I, p. 329.

of a necessary relation between things: it can know the form of that relation only by experience.

Two things must, therefore, be considered in the experimental method: (1) The art of getting accurate facts by means of rigorous investigation; (2) the art of working them up by means of experimental reasoning, so as to deduce knowledge of the law of phenomena. We said that experimental reasoning always and necessarily deals with two facts at a time: observation, used as a starting point; experiment, used as conclusion or control. In reasoning, however, we can distinguish between actual observation and experiment only, as it were, by logical abstraction and because of the position in which they stand.

But outside of experimental reasoning, observation and experiment no longer exist in this abstract sense; there are only concrete facts in each, to be got by precise and rigorous methods of investigation. We shall see, further on, that the investigator himself must be analyzed into observer and experimenter; not according to whether he is active or passive in producing phenomena, but according to whether he acts on them or not, to make himself their master.

III. THE INVESTIGATOR; SCIENTIFIC RESEARCH

The art of investigation is the cornerstone of all the experimental sciences. If the facts used as a basis for reasoning are ill-established or erroneous, everything will crumble or be falsified; and it is thus that errors in scientific theories most often originate in errors of fact.

In investigation, considered as the art of experimental research, we find only facts brought to light by investigators and noted as rigorously as possible with the help of the most suitable means. There is no further occasion here to distinguish observers from experimenters by the character of the processes of investigation used. In the last section I showed that the definitions and distinctions which men have tried to set up on the basis of the investigator's activity or passivity cannot be sustained. Observers and experimenters, indeed, are investigators seeking to note facts to the best of their ability, using more or less complicated means for this purpose according to the complexity of the phenomena that they study. Both need the same manual and intellectual activity, the same dexterity, the same spirit of invention, to create and perfect the different pieces of apparatus or instruments for investigation which, for the most part, they have in common. Every science has its own kind of investigation and its equipment of special instruments and methods. This, after all, is plain enough, since every science is characterized by the nature of its problems and by the variety of the phenomena that it studies. Medical investigation is the most complicated of all: it includes all the methods proper to anatomical, physiological and therapeutic research, and, as it develops, it also borrows from chemistry and physics many means of research which become powerful allies. In the experimental sciences all progress is measured by improvement in the means of investigation. The whole future of experimental medicine depends on creating a method of research which may be applied fruitfully to the study of vital phenomena, whether in a normal or abnormal state. I shall not here dwell on the necessity of such a method of investigation in experimental medicine, and I shall not even attempt to enumerate the difficulties. I shall limit myself to saying that my whole scientific life is devoted to contributing my share to the immense work which modern science will have the glory of having understood, and the merit of having begun, while leaving to future ages the task of continuing and finally establishing it. The two volumes which will form my work on the Principles of Experimental Medicine will be devoted solely to elaborating the methods of experimental investigation applied to physiology, pathology and therapeutics. But as no one man can consider all aspects of medical investigation, I shall limit myself further in this vast subject, by dealing especially with systematization of the methods of zoölogical vivisection. It cannot be gainsaid that this is the most delicate and difficult branch of biological investigation; but I deem it the most fruitful and perhaps the most immediately useful for the advancement of experimental medicine.

In scientific investigation, minutiae of method are of the highest importance. The happy choice of an animal, an instrument constructed in some special way, one reagent used instead of another, may often suffice to solve the most abstract and lofty questions. Every time that a new and reliable means of experimental analysis makes its appearance, we invariably see science make progress in the questions to which this means of analysis can be applied. On the contrary, a bad method or defective processes of research may cause the gravest errors, and may retard science by leading it astray. In a word, the greatest scientific truths are rooted in details of experimental investigation which form, as it were, the soil in which these truths develop.

One must be brought up in laboratories and live in them, to appreciate the full importance of all the details of procedure in investigation, which are so often neglected or despised by the false men of science calling themselves generalizers. Yet we shall reach really fruitful and luminous generalizations about vital phenomena only in so far as we ourselves experiment and, in hospitals, amphitheatres, or laboratories, stir the fetid or throbbing ground of life. It has somewhere been said that true science is like a flowering and delectable plateau which can be attained only after climbing craggy steeps and scratching one's legs against branches and brushwood. If a comparison were required to express my idea of the science of life, I should say that it is a superb and dazzlingly lighted hall which may be reached only by passing through a long and ghastly kitchen.

IV. OBSERVERS AND EXPERIMENTERS; THE SCIENCES OF OBSERVATION AND OF EXPERIMENT

We have just seen that, from the point of view of the art of investigation, observation and experiment should be considered only as *facts* brought out by investigators, and we have added that methods of investigation do not differentiate the men who observe from the men who experiment. Where then, you will ask, is the difference between observers and experimenters? It is here: we give the name observer to the man who applies methods of investigation, whether simple or complex, to the study of phenomena which he does not vary and which he therefore gathers as nature offers them. We give the name experimenter to the man who applies methods of investigation, whether simple or complex, so as to make natural phenomena vary, or so as to alter them with some purpose or other, and to make them present themselves in circumstances or conditions in which nature does not show them. In this sense, observation is investigation of a natural phenomenon, and experiment is investigation of a phenomenon altered by the investigator. We shall see that this distinction, apparently quite external and depending simply on a

observation is.

definition of words, still supplies the one meaning with which to grasp the important difference separating sciences of observation from sciences of experimentation or experimental sciences. [We said, in an earlier paragraph, that the words observation and experiment, taken in an abstract sense, mean, the first, purely and simply noting a fact, the second, testing an idea by a fact. But if we consider observation merely in this abstract sense, we cannot deduce from it any science of observation. By simply noting facts, we can never succeed in establishing a science. Pile up facts or observations as we may, we shall be none the wiser. To learn, we must necessarily reason about what we have observed, compare the facts and judge them by other facts used as controls. But one observation may serve as control for another observation, so that a science of observation is simply a science made up of observations, i.e., a science in which we reason about facts observed in their natural state, as we have already defined them. An experimental science, or science of experimentation, is a science made up of experiments, i.e., one in which we reason on experimental facts found in conditions created and determined by the experimenter himself.]

Certain sciences, like astronomy, will always remain sciences of observation, because the phenomena studied are outside our sphere of action; but terrestrial sciences may be, at once, sciences of observation and experimental sciences. Let me add that all these sciences begin as sciences of pure observation; only as we go into the analysis of phenomena do they become experimental, because the observer, turning experimenter, invents methods of investigation to penetrate bodies and vary the conditions of phenomena. Experimentation is only utilizing methods of investigation peculiar to experimenters.

[Now experimental reasoning is absolutely the same, whether in sciences of observation or in experimental sciences. We find the same judgment by comparison based on two facts, one used as starting point, the other as conclusion, of our reasoning. Only in the sciences of observation, the two facts are always observations; while in the experimental sciences, the two facts may be taken exclusively from experimentation, or at the same time from experimentation and from observation, according to the special case and according to how deeply we go into experimental analysis. A physician observing a disease in different circumstances, reasoning about the influence of these circumstances, and deducing consequences which are controlled by other observations,—this physician reasons experimentally, even though he makes no experiments. But if he wishes to go further, and to know the inner mechanism of the disease, he will have to deal with hidden phenomena, and so he will experiment; but he will still reason in the same way.

A naturalist observing animals in all the conditions necessary to their existence, and deducing from these observations consequences verified and controlled by other observations,—such a naturalist uses the experimental method even though he performs no experiments, properly speaking. But if he has to go on to observe phenomena inside the stomach, he is forced to invent more or less complex methods of experimentation in order to look inside a cavity hidden from sight. His experimental reasoning, nevertheless, is the same; Réaumur and Spallanzani alike apply the experimental method when making their observations of natural history or their experiments with digestion. When Pascal made a barometric observation at the bottom of the Tour Saint Jacques, and later took another at the top of the tower, we must admit that he performed an experiment; yet here were simply two comparative observations of air pressure carried out in view of the preconceived idea that this pressure should vary according to height. On the other hand, when Jenner,[5] in observing a cuckoo on a tree, used a spy-glass so as not to frighten it, he made a mere observation, because he did not compare this cuckoo with a previous cuckoo, to deduce a conclusion from the observation and to form a judgment about it. In the same way an astronomer first makes observations and then reasons about them to deduce a system of ideas which he controls by observations made in conditions suited to his purpose. The astronomer reasons like an experimenter, because the experience which he gains implies judgment throughout and comparison between two facts bound together in the mind by an idea.

However, as we have said already, we must clearly differentiate astronomers from the men of science concerned with terrestrial science, in that astronomers limit themselves perforce to observation, as they cannot go into the skies to experiment on the planets. In this power of the investigator to act on phenomena, precisely here

[5] Jenner, *On the natural history of the cuckoo.* (*Philosophical Transactions,* 1788, Chap. XVI, p. 432.)

is the difference separating the so-called sciences of experimentation from those of observation.

Laplace considers astronomy a science of observation, because we can only observe the movements of the planets; we cannot reach them, indeed, to alter their course and to experiment with them. "On earth," said Laplace, "we make phenomena vary by experiments; in the sky, we carefully define all the phenomena presented to us by celestial motion." [6] Certain physicians call medicine a science of observations, because they wrongly think that experimentation is inapplicable to it.

Fundamentally, all sciences reason in the same way and aim at the same object. They all try to reach knowledge of the law of phenomena, so as to foresee, vary or master phenomena. Astronomers foretell the movements of the stars; they deduce from them a quantity of practical ideas; but they cannot alter celestial phenomena by experimentation as do chemists and physicists the phenomena of their sciences.

If then, from the point of view of philosophic method, there is no essential difference between sciences of observation and sciences of experimentation, still there is a real one from the point of view of the practical consequences, which man deduces from them, and the power which he gains by their means. In sciences of observation, man observes and reasons experimentally, but he does not experiment; and in this sense we might say that a science of observation is a passive science. In sciences of experimentation, man observes, but in addition he acts on matter, analyzes its properties and to his own advantage brings about the appearance of phenomena which doubtless always occur according to natural laws, but in conditions which nature often has not yet achieved. With the help of these active experimental sciences, man becomes an inventor of phenomena, a real foreman of creation; and under this head we cannot set limits to the power that he may gain over nature through future progress in the experimental sciences.

The question remains whether medicine should continue a science of observation or become an experimental science. Medicine must doubtless begin as simple clinical observation. Then, since the human organism is in itself a harmonious unit, a little world (microcosm) contained in the great world (macrocosm), men have actually

[6] Laplace. *Système du monde.* Chap. II.

maintained that life is indivisible and that we should limit ourselves to observing the phenomena presented to us as a whole by living organisms, whether well or sick, and should content ourselves
with reasoning on the facts observed. But if we admit that we must
so limit ourselves, and if we posit as a principle that medicine is
only a passive science of observation, then physicians should no more
touch the human body than astronomers touch the planets. Hence,
normal and pathological anatomy, vivisection applied to physiology,
pathology and therapeutics,—all would become completely useless.
Medicine so conceived can lead only to prognosis and to hygienic
prescriptions of doubtful utility; it is the negation of active medicine, i.e., of real and scientific therapeutics.

This is by no means the place to begin examining so important
a definition as that of experimental medicine. I propose to treat
this question later with all necessary amplification. I shall limit
myself here to saying that I think that medicine is destined to be
an experimental and progressive science; and precisely because of
my conviction in this respect, I am putting together this work
with the object of contributing my share toward encouraging the
development of scientific and experimental medicine.

V. Experiment Is Fundamentally Only Induced Observation

Despite the important difference, which we have just pointed out,
between the so-called sciences of observation and of experimentation,
observers and experimenters still have the common and immediate
object, in their investigations, of establishing and noting facts and
phenomena as rigorously as possible, and with the help of the most
appropriate means; they behave exactly as if they were dealing with
two ordinary observations. In both cases, indeed, a fact is simply
noted; the only difference is this,—as the fact which an experimenter
must verify does not present itself to him naturally, he must make
it appear, i.e., induce it, for a special reason and with a definite object. Hence we may say that an experiment is fundamentally just an
observation induced with some object or other. In the experimental
method, search for facts, i.e., investigation, is always accompanied by
reasoning, so that experimenters usually make an experiment to
control or verify the value of an experimental idea. Hence, in this

*ideas & needed before experimentation
begins*

case, the experiment is an observation induced with the object of control.]

Still, to complete our definition and to extend it to the sciences of observation, it is worth recalling here that, to verify an idea, it is not always absolutely necessary to make an experiment or an observation ourselves. We shall have recourse to experimentation perforce only when the observation to be induced is not already prepared in nature. But if an observation has already been made, either naturally or accidentally, or even by another investigator, then we may take it ready made, and produce it simply to serve as verification of the experimental idea. [And this may be summed up again by saying that, in this case, the experiment is just an observation produced for the purpose of control. It follows that, to reason experimentally, we must usually have an idea and afterwards induce or produce facts, i.e., observations, to control our preconceived idea.]

We shall examine later the importance of preconceived experimental ideas; let it suffice us now to say that the idea, by virtue of which we undertake an experiment, may be more or less clearly defined, according to the nature of the subject and according to the state of perfection of the science in which we are experimenting. Indeed the guiding idea of an experiment should include everything already known about the subject, so as to direct our search more surely toward problems whose solution may be fruitful in the advancement of science. [In established sciences, like physics and chemistry, experimental ideas are deduced in logical sequence from ruling theories, and are submitted with a clearly defined meaning to the control of experiment; [but in the case of a science in its infancy, like medicine, where complex and obscure questions are still to be studied, experimental ideas do not always emerge from rather vague conceptions. What then must be done? Must we abstain and wait for observations to present themselves spontaneously and so bring us clearer ideas? We might often wait long and even in vain; in any case we gain by experimenting. But in this instance we can guide ourselves only by a kind of intuition, as we catch sight of probabilities; and if the subject is entirely dark and unexplored, physiologists should not be afraid even to act somewhat at random, so as to try,—permit me the common expression,—fishing in troubled waters. This amounts to saying that, in the midst of the functional disturbances which they produce, they may hope to see some unex-

various phases —

pected phenomena emerge which may give direction to their research. Such groping experiments, which are very common in physiology and therapeutics because of the complex and backward state of these sciences, may be called *experiments to see,* because they are intended to make a first observation emerge, unforeseen and undetermined in advance, but whose appearance may suggest an experimental idea and open a path for research.

There are instances, then, in which we experiment without having a probable idea to verify. However, experimentation in this instance is none the less intended to induce an observation, only it induces it with a view to finding an idea which shall point out a later path to follow in investigation. We may therefore say that the experiment is then an observation induced with the object of bringing to birth an idea.

To sum up, the investigator seeks and concludes; he includes both observations and experiments, he pursues the discovery of new ideas, even while seeking facts from which to draw a conclusion, or an experiment calculated to control other ideas.

In a general and abstract sense, an experimenter, then, is a man who produces or induces, in definite conditions, observed facts, to derive from them the instruction which he wishes,—that is, experience. An observer is a man who gathers observed facts and who decides whether they have been ascertained by the help of appropriate means. Thus it is that experimenters must at the same time be good observers, and that in the experimental method, experiment and observation always advance side by side.

VI. IN EXPERIMENTAL REASONING, EXPERIMENTERS ARE NOT SEPARATE FROM OBSERVERS

Men of science who mean to embrace the principles of the experimental method as a whole, must fulfill two classes of conditions and must possess two qualities of mind which are indispensable if they are to reach their goal and succeed in the discovery of truth. First, they must have ideas which they submit to the control of facts; but at the same time they must make sure that the facts which serve as starting point or as control for the idea are correct and well established; they must be at once observers and experimenters.

Observers, we said, purely and simply note the phenomena before their eyes. They must be anxious only to forearm themselves against errors of observation which might make them incompletely see or poorly define a phenomenon. To this end they use every instrument which may help make their observations more complete. Observers, then, must be photographers of phenomena; their observations must accurately represent nature. We must observe without any preconceived idea; the observer's mind must be passive, that is, must hold its peace; it listens to nature and writes at nature's dictation.

But when a fact is once noted and a phenomenon well observed, reasoning intervenes, and the experimenter steps forward to interpret the phenomenon.

An experimenter, as we have already said, is a man inspired by a more or less probable but anticipated interpretation of observed phenomena, to devise experiments which, in the logical order of his anticipations, shall bring results serving as controls for his hypothesis or preconceived idea. To do this, an experimenter reflects, tries out, gropes, compares, contrives, so as to find the experimental conditions best suited to gain the end which he sets before him. Of necessity we experiment with a preconceived idea. An experimenter's mind must be active, i.e., must question nature, and put all manner of queries to it according to the various hypotheses which suggest themselves.

But when the conditions of an experiment are once established and worked up according to the mind's preconceived idea, an induced or premeditated observation will, as we said, result. Phenomena then appear which the experimenter has caused, but which must now be noted, so as to learn next how to use them to control the experimental idea which brought them to birth. Now, from the moment when the result of an experiment appears, the experimenter is confronted with a real observation which he has induced and must note, like any other observation, without any preconceived idea. The experimenter must now disappear or rather change himself instantly into an observer; and it is only after he has noted the results of the experiment exactly, like those of an ordinary observation, that his mind will come back, to reason, compare and decide whether his experimental hypothesis is verified or disproved by these very results. To maintain the comparison suggested above, I may say that our experimenter puts questions to nature; but that, as soon as she speaks, he must hold his peace; he must note her answer, hear her out and in every case accept her decision. It has been said that the experimenter must force nature to unveil herself. Yes, the experimenter doubtless forces nature to unveil herself by attacking her with all manner of questions; he must never answer for her nor listen partially to her answers by taking, from the results of an experiment, only those which support or confirm his hypothesis. We shall see later that this is one of the great stumbling blocks of the experimental method. An experimenter, who clings to his preconceived idea and notes the results of his experiment only from this point of view, falls inevitably into error, because he fails to note what he has not foreseen and so makes a partial observation. An experimenter must not hold to his idea, except as a means of inviting an answer from nature. But he must submit his idea to nature and be ready to abandon, to alter or to supplant it, in accordance with what he learns from observing the phenomena which he has induced.

Two operations must therefore be considered in any experiment. The first consists in premeditating and bringing to pass the conditions of the experiment; the second consists in noting the results of the experiment. It is impossible to devise an experiment without a preconceived idea; devising an experiment, we said, is putting a question; we never conceive a question without an idea which invites an answer. I consider it, therefore, an absolute principle that experiments must always be devised in view of a preconceived idea, no matter if the idea be not very clear nor very well defined. As for noting the results of the experiment, which is itself only an induced observation, I posit it similarly as a principle that we must here, as always, observe without a preconceived idea.

In the experimenter, we might also differentiate and separate the man who preconceives and devises an experiment from the man who carries it out or notes its results. In the former, it is the scientific investigator's mind that acts; in the latter, it is the senses that observe and note. What I am setting forth is most strikingly proved in the case of François Huber.[7] Though blind, this great naturalist left us admirable experiments which he conceived and afterward had carried out by his serving man, who, for his part, had not a single scientific idea. So Huber was the directing mind that

[7] François Huber, *Nouvelles Observations sur les Abeilles*, 2nd edition, expanded by his son, Pierre Huber, Geneva, 1814.

devised the experiment; but he was forced to borrow another's senses. The serving man stood for the passive senses, obedient to the mind in carrying out an experiment devised in the light of a preconceived idea.

People who condemn the use of hypotheses and of preconceived ideas in the experimental method make the mistake of confusing invention of an experiment with noting its results. We may truly say that the results of an experiment must be noted by a mind stripped of hypotheses and preconceived ideas. But we must beware of proscribing the use of hypotheses and of ideas when devising experiments or imagining means of observation. On the contrary, as we shall soon see, we must give free rein to our imagination; the idea is the essence of all reasoning and all invention. All progress depends on that. It cannot be smothered or driven away on the pretence that it may do harm; it must only be regulated and given a criterion, which is quite another matter.

The true scientist is one whose work includes both experimental theory and experimental practice. (1) He notes a fact; (2) à propos of this fact, an idea is born in his mind; (3) in the light of this idea, he reasons, devises an experiment, imagines and brings to pass its material conditions; (4) from this experiment, new phenomena result which must be observed, and so on and so forth. The mind of a scientist is always placed, as it were, between two observations: one which serves as starting point for reasoning, and the other which serves as conclusion.

To make myself clearer, I have endeavored to separate the different operations of experimental reasoning. But when it all takes place at the same time in the head of a scientist, abandoning himself to investigation in a science as vague as medicine still is, then the results of observation are so entangled with the bases of experiment that it would be alike impossible and useless to try to dissociate, from their inextricable mingling, each one of these terms. It is enough to remember the principle that an a priori idea, or better, an hypothesis, is a stimulus to experiment, and that we must let ourselves go with it freely, provided that we observe the results of our experiment rigorously and fully. If an hypothesis is not verified and disappears, the facts which it has enabled us to find are none the less acquired as indestructible materials for science.

Observers and experimenters, then, correspond to different phases of experimental research. The observer does not reason, he notes; the experimenter, on the other hand, reasons and grounds himself on acquired facts, to imagine and induce rationally other facts. But though in theory and abstractly we may differentiate observers from experimenters, it seems impossible to separate them in practice, since we see that one and the same investigator, perforce, is alternately observer and experimenter.

Things happen constantly, indeed, in this way when a single man of science discovers and explains a whole scientific question unaided. But it more often happens in the evolution of science, that different parts of experimental reasoning are shared by several men. Some of these, both in medicine and in natural history, merely gather and assemble observations; others manage to formulate more or less ingenious and more or less probable hypotheses based on these observations; then others come in to create conditions favoring the birth of an experiment to control these hypotheses; finally others apply themselves more especially to generalizing and systematizing the results obtained by the different observers and experimenters. This parceling out of the experimental domain is useful, because each one of its various parts is all the better cultivated. In fact we can easily conceive that, in certain sciences, the means of observation and experimentation are such specialized instruments that their management and use require a certain manual dexterity or the sharpening of certain senses. But while I accept specialization in the practice, I reject it utterly in the theory of science. I believe, indeed, that making generalization one's specialty is anti-philosophic and anti-scientific, in spite of what has been proclaimed by a modern philosophic school which piques itself on its scientific basis.

Experimental science, however, cannot advance on a single side of the method taken separately; it goes ahead only by the union of all parts of the method converging toward a common goal. Men who gather observations are useful only because their observations are afterward introduced into experimental reasoning; in other words, endless accumulation of observations leads nowhere. Men, who formulate hypotheses à *propos* of observations gathered by others, are useful only in so far as men seek to verify these hypotheses by experimenting; else these hypotheses, unverified or unverifiable by experiment, would engender nothing but systems and would bring us back to scholasticism. Men who experiment, despite all their dexterity, cannot solve problems unless they are inspired by a fortunate hypothesis based on accurate and well-made observations. Finally men who generalize can make lasting theories only in so far as they themselves learn all the scientific details that these theories are intended to represent. Scientific generalization must proceed from particular facts to principles; and principles are the more stable as they rest on deeper details, just as a stake is the firmer, the farther it is driven into the ground.

We see, then, that the elements of the scientific method are interrelated. Facts are necessary materials; but their working up by experimental reasoning, i.e., by theory, is what establishes and really builds up science. Ideas, given form by facts, embody science. A scientific hypothesis is merely a scientific idea, preconceived or previsioned. A theory is merely a scientific idea controlled by experiment. Reasoning merely gives a form to our ideas, so that everything, first and last, leads back to an idea. The idea is what establishes, as we shall see, the starting point or the *primum movens* of all scientific reasoning, and it is also the goal in the mind's aspiration toward the unknown.

CHAPTER II

THE A PRIORI IDEA AND DOUBT IN EXPERIMENTAL REASONING

Everyone first works out his own ideas about what he sees and is inclined to interpret natural phenomena by anticipation before knowing them through experience. This tendency is spontaneous; a preconceived idea always has been and always will be the first flight of an investigating mind. But the object of the experimental method is to transform this *a priori* conception, based on an intuition or a vague feeling about the nature of things, into an *a posteriori* interpretation founded on the experimental study of phenomena. This is why the experimental method is also called the *a posteriori* method.

Man is by nature metaphysical and proud. He has gone so far as to think that the idealistic creations of his mind, which correspond to his feelings, also represent reality. Hence it follows that the experimental method is by no means primitive or natural to man, and that only after lengthy wanderings in theological and scholastic discussion has he recognized at last the sterility of his efforts in this direction. At this point man becomes aware that he cannot dictate laws to nature, because he does not contain within himself the knowledge and criterion of external things, and he understands that to find truth he must, on the contrary, study natural laws and submit his ideas, if not his reason, to experience, that is, to the criterion of facts. Yet for all that, the method of work of the human mind is not changed at bottom. The metaphysician, the scholastic, and the experimenter all work with an *a priori* idea. The difference is that the scholastic imposes his idea as an absolute truth which he has found, and from which he then deduces consequences by logic alone. The more modest experimenter, on the other hand, states an idea as a question, as an interpretative, more or less probable anticipation of nature, from which he logically deduces consequences which, moment by moment, he confronts with reality by means of experiment. He advances, thus, from partial to more general truths, but without ever daring to assert that he has grasped the absolute truth.

Indeed if we held it at any point whatever, we should have it everywhere; for the absolute leaves nothing outside itself.

An experimental idea, then, is also an *a priori* idea, but it is an idea that presents itself in the form of an hypothesis the consequences of which must be submitted to the criterion of experiment, so that its value may be tested. The experimenter's mind differs from the metaphysician's or the scholastic's in its modesty, because experiment makes him, moment by moment, conscious of both his relative and his absolute ignorance. In teaching man, experimental science results in lessening his pride more and more by proving to him every day that primary causes, like the objective reality of things, will be hidden from him forever and that he can know only relations. Here is, indeed, the one goal of all the sciences, as we shall see further on.

The human mind has at different periods of its evolution passed successively through *feeling, reason* and *experiment*. First, feeling alone, imposing itself on reason, created the truths of faith or theology. Reason or philosophy, the mind's next mistress, brought to birth scholasticism. At last, experiment, or the study of natural phenomena, taught man that the truths of the outer world are to be found ready formulated neither in feeling nor in reason. These are indispensable merely as guides; but to attain external truths we must of necessity go down into the objective reality of things where they lie hidden in their phenomenal form.

Thus, in the natural progress of things, appeared the experimental method which includes everything and which, as we shall soon see, leans successively on the three divisions of that unchangeable tripod: sentiment, reason and experiment. In the search for truth by means of this method, feeling always takes the lead, it begets the *a priori* idea or intuition; reason or reasoning develops the idea and deduces its logical consequences. But if feeling must be clarified by the light of reason, reason in turn must be guided by experiment.

I. Experimental Truths Are Objective or External

The experimental method is concerned only with the search for objective truths, not with any search for subjective truths.

As there are two kinds of functions in man's body, the first, conscious functions, the rest not, so in his mind there are two kinds of truths or notions, some conscious, inner or subjective, the others unconscious, outer or objective. Subjective truths are those flowing from principles of which the mind is conscious, and which bring it the sensation of absolute and necessary evidence. The greatest truths, indeed, are at bottom simply a feeling in our mind; that is what Descartes meant by his famous aphorism.

We said, on the other hand, that man would never know either the primary cause, nor the essence of things. Hence truth never shows itself to his mind except in the form of a connection or of a necessary and absolute relation. But this connection may be absolute only in so far as its conditions are simple and subjective, that is, when the mind is aware of knowing them all. Mathematics embodies the relations of things in conditions of ideal simplicity. It follows that these principles or relations, once found, are accepted by the mind as absolute truths, i.e., truths independent of reality. We see now that all logical deductions in a piece of mathematical reasoning are just as certain as their principle, and that they do not require verification by experiment. That would be trying to place the senses above reason; and it would be absurd to seek to prove what is absolutely true for the mind and what it could not conceive otherwise.

But when man stops working with subjective relations, the conditions of which his mind has created, and tries to learn about the objective relations of nature which he has not created, then at once the inner and conscious criterion fails him. He is, of course, still aware that in the objective or outer world truth consists, in the same way, of necessary relations; but he lacks knowledge of the conditions of these relations. Only if he had created these conditions, indeed, could he possess absolute knowledge of them and absolute understanding.

Still man must believe that the objective relations between phenomena of the outer world might attain the certainty of subjective truths if they were reduced to a state of simplicity that his mind could completely grasp. Thus, in the study of the simplest of natural phenomena, experimental science has laid hold on certain relations which appear absolute. Such are the propositions which serve as principles in theoretical mechanics and in some branches of mathematical physics. In these sciences, indeed, we reason by logical deduction which we do not submit to experiment, because we admit, as in mathematics, that the principle being true the deductions are true also.   Still, there is a wide difference to note in this respect, that the starting point here is no longer a subjective and conscious truth, but an objective and unconscious truth, borrowed from observation or experiment.   Now this truth is never more than relative to the number of experiments and observations that have been made. Even if no observation has so far disproved the truth in question, still the mind does not therefore imagine that things cannot happen otherwise; so that it is only by hypothesis that we admit the principle as absolute.   That is why the application of mathematical analysis to natural phenomena, even very simple ones, may have its dangers if experimental verification is entirely rejected.   In this case, mathematical analysis becomes a blind instrument, if we do not from time to time retemper it in the furnace of experiment.   I here express a thought uttered by many great mathematicians and great physicists; and in order to recall one of the most authoritative opinions in this field, I will cite what my learned associate and friend, J. Bertrand, wrote on this subject in his fine tribute to Sénarmont: "For the physicist, geometry should be only a powerful ally: when it has pushed its principles to their last consequences, it can do no more, and the uncertainty of the starting point can only be increased by the blind logic of analysis, if experiment at each step does not serve as compass and ruler." [1]

Theoretical mechanics and mathematical physics make the connection then between mathematics proper and the experimental sciences. They include the simplest cases.   But as soon as we go into physics and chemistry, and especially biology, the phenomena are complicated by so many relations that the principles, embodied in theories to which we have been able to rise, are only provisional and are so hypothetical that our deductions, even though very logical, are absolutely uncertain and can in no case dispense with experimental verification.

In short, man may relate all his reasonings to two criteria: the one, inner and conscious, is sure and absolute; the other, outer and unconscious, is experimental and relative.

When we reason about outer objects, but consider them in their relation to ourselves according to the pleasure or displeasure which they occasion us in proportion to their utility or their disadvantages,

---

[1] J. Bertrand, *Eloge de M. Sénarmont*, address given at the sixth annual public meeting of the *Société de secours des amis des sciences*.

we still possess an inner criterion in our sensations. So, when we reason about our own actions, we again have a sure guide, because we are conscious of what we are thinking and of what we are feeling. But if we wish to judge the actions of another man and to know the motives which make him act, then it is quite different. Doubtless we see before our eyes the man's movements and the acts which, we are sure, are expressions of his feeling and his will. What is more, we also admit that there is a necessary relation between actions and their cause; but what is this cause? We do not feel it ourselves, we are not aware of it as in our own case; we are therefore forced to interpret and imagine it from the movements that we see and the words that we hear. So we must verify the man's acts, one by another; we consider how he behaves in such and such circumstances, and in short, we turn to the experimental method. In like manner, when a man of science considers the natural phenomena which surround him and which he wishes to know in themselves and in their complex mutual relations of causation, every inner criterion fails him, and he is forced to invoke experiment to verify the suppositions and the reasonings that he is making about them. Experiment, then, according to Goethe's expression, becomes the one mediator between the objective and the subjective,[2] that is to say, between the man of science and the phenomena which surround him.

Experimental reasoning is the only reasoning that naturalists and physicians can use in seeking the truth and approaching it as nearly as possible. Indeed, in its very character as an outer and unconscious criterion, experiment gives only relative truth, without being able to prove to the mind that it knows truth absolutely.

An experimenter facing natural phenomena is like a spectator watching a dumb show. He is in some sort the examining magistrate for nature; only instead of grappling with men who seek to deceive him by lying confessions or false witness, he is dealing with natural phenomena which for him are persons whose language and customs he does not know, persons living in the midst of circumstances unknown to him, yet persons whose designs he wishes to learn. For this purpose he uses all the means within his power. He observes their actions, their gait, their behavior, and he seeks to disengage their cause by means of various attempts, called experiments. He

[2] Goethe, _Œuvres d'histoire naturelle_, translation by M. Ch. Martins, Introduction, p. 1.

uses every imaginable artifice, and, as the popular expression goes, he often makes a false plea in order to learn the truth. In all this, the experimenter reasons necessarily according to his own character and lends to nature his own ideas. He makes suppositions about the cause of actions taking place before his eyes; and to learn whether the hypothesis which serves as groundwork for his interpretation is correct, he takes measures to make facts appear which in the realm of logic may be either the confirmation or the negation of the idea which he has conceived. Now, I repeat, only this logical verification can teach him and give him *experience*. The naturalist observing animals whose behavior and habits he wishes to know, the physiologist and the physician wishing to study the hidden functions of living bodies, the physicist and the chemist defining the phenomena of inert matter,—they are all in the same situation; they have manifestations before them which they can interpret only with the help of the experimental criterion, the only one which we need to consider here.

II. Intuition or Feeling Begets the Experimental Idea

We said above that the experimental method rests successively on feeling, reason and experiment.

Feeling gives rise to the experimental idea or hypothesis, i.e., the previsioned interpretation of natural phenomena. The whole experimental enterprise comes from the idea, for this it is which induces experiment. Reason or reasoning serves only to deduce the consequences of this idea and to submit them to experiment.

An anticipative idea or an hypothesis is, then, the necessary starting point for all experimental reasoning. Without it, we could not make any investigation at all nor learn anything; we could only pile up sterile observations. If we experimented without a preconceived idea, we should move at random, but, on the other hand, as we have said elsewhere, if we observed with preconceived ideas, we should make bad observations and should risk taking our mental conceptions for reality.

Experimental ideas are by no means innate. They do not arise spontaneously; they must have an outer occasion or stimulant, as is the case in all physiological functions. To have our first idea of things, we must see those things; to have an idea about a natural phenomenon, we must, first of all, observe it. The mind of man cannot conceive an effect without a cause, so that the sight of a phenomenon always awakens an idea of causation. All human knowledge is limited to working back from observed effects to their cause. Following an observation, an idea connected with the cause of the observed phenomenon presents itself to the mind. We then inject this anticipative idea into a train of reasoning, by virtue of which we make experiments to control it.

Experimental ideas, as we shall later see, may arise either *a priori* of a fact observed by chance or following some experimental venture or as corollaries of an accepted theory. For the moment, we may merely note that the experimental idea is by no means arbitrary or purely imaginative; it must always have a support in observed reality, that is to say, in nature. The experimental hypothesis, in short, must always be based on prior observation. Another essential of any hypothesis is that it must be as probable as may be and must be experimentally verifiable. Indeed if we made an hypothesis which experiment could not verify, in that very act we should leave the experimental method to fall into the errors of the scholastics and makers of systems.

À propos of a given observation, no rules can be given for bringing to birth in the brain a correct and fertile idea that may be a sort of intuitive anticipation of successful research. The idea once set forth, we can only explain how to submit it to the definite precepts and precise rules of logic from which no experimenter may depart; but its appearance is wholly spontaneous, and its nature is wholly individual. A particular feeling, a *quid proprium* constitutes the originality, the inventiveness, or the genius of each man. A new idea appears as a new or unexpected relation which the mind perceives among things. All intellects doubtless resemble each other, and in all men similar ideas may arise in the presence of certain simple relations between things, which everyone can grasp. But like the senses, intellects do not all have the same power or the same acuteness; and subtle and delicate relations exist which can be felt, grasped and unveiled only by minds more perceptive, better endowed, or placed in intellectual surroundings which predispose them favorably.

If facts necessarily gave birth to ideas, every new fact ought to beget a new idea. True, this is what most often takes place; for new facts exist, the character of which makes the same new idea come to all men, placed in the same circumstances as respects previous information. But facts also exist which mean nothing to most minds, while they are full of light for others. It even happens that a fact or an observation stays a very long time under the eyes of a man of science without in any way inspiring him; then suddenly there comes a ray of light, and the mind interprets the fact quite differently and finds for it wholly new relations. The new idea appears, then, with the rapidity of lightning, as a kind of sudden revelation; which surely proves that in this case the discovery inheres in a feeling about things which is not only individual, but which is even connected with a transient condition of the mind. The experimental method, then, cannot give new and fruitful ideas to men who have none; it can serve only to guide the ideas of men who have them, to direct their ideas and to develop them so as to get the best possible results. The idea is a seed; the method is the earth furnishing the conditions in which it may develop, flourish and give the best of fruit according to its nature. But as only what has been sown in the ground will ever grow in it, so nothing will be developed by the experimental method except the ideas submitted to it. The method itself gives birth to nothing. Certain philosophers have made the mistake of according too much power to method along these lines.

The experimental idea is the result of a sort of presentiment of the mind which thinks things will happen in a certain way. In this connection we may say that we have in our minds an intuition or feeling as to the laws of nature, but we do not know their form. We can learn it only from experiment.

Men with a presentiment of new truths are rare in all the sciences; most men develop and follow the ideas of a few others. Those who make discoveries are the promoters of new and fruitful ideas. We usually give the name of discovery to recognition of a new fact; but I think that the idea connected with the discovered fact is what really constitutes the discovery. Facts are neither great nor small in themselves. A great discovery is a fact whose appearance in science gives rise to shining ideas, whose light dispels many obscurities and shows us new paths. There are other facts which, though new, teach us but little; they are therefore small discoveries. Finally, there are new facts which, though well observed, teach nothing to anyone; they remain, for the moment, detached and sterile in science; they are what we may call raw facts or crude facts.

Discovery, then, is a new idea emerging in connection with a fact found by chance or otherwise. Consequently, there can be no method for making discoveries, because philosophic theories can no more give inventive spirit and aptness of mind to men, who do not possess them, than knowledge of the laws of acoustics or optics can give a correct ear or good sight to men deprived of them by nature. But good methods can teach us to develop and use to better purpose the faculties with which nature has endowed us, while poor methods may prevent us from turning them to good account. Thus the genius of inventiveness, so precious in the sciences, may be diminished or even smothered by a poor method, while a good method may increase and develop it. In short, a good method promotes scientific development and forewarns men of science against those numberless sources of error which they meet in the search for truth; this is the only possible object of the experimental method. In biological science, the rôle of method is even more important than in other sciences, because of the immense complexity of the phenomena and the countless sources of error which complexity brings into experimentation. Yet even from the biological point of view, we cannot claim to treat the experimental method completely here; we must limit ourselves to giving a few general principles for the guidance of minds applying themselves to research in experimental medicine.

III. Experimenters Must Doubt, Avoid Fixed Ideas, and Always Keep Their Freedom of Mind

The first condition to be fulfilled by men of science, applying themselves to the investigation of natural phenomena, is to maintain absolute freedom of mind, based on philosophic doubt. Yet we must not be in the least sceptical; we must believe in science, i.e., in determinism; we must believe in a complete and necessary relation between things, among the phenomena proper to living beings as well as in all others; but at the same time we must be thoroughly convinced that we know this relation only in a more or less approximate way, and that the theories we hold are far from embodying changeless truths. When we propound a general theory in our sciences, we are sure only that, literally speaking, all such theories are false. They are only partial and provisional truths which are necessary to us, as steps on which we rest, so as to go on with investigation; they embody only the present state of our knowledge, and consequently they must change with the growth of science, and all the more often when sciences are less advanced in their evolution. On the other hand, our ideas come to us, as we said, in view of facts which have been previously observed and which we interpret afterward. Now countless sources of error may slip into our observations, and in spite of all our attention and sagacity, we are never sure of having seen everything, because our means of observation are often too imperfect. The result of all this is, then, that if reasoning guides us in experimental science, it does not necessarily force its deductions upon us. Our mind can always remain free to accept or to dispute these deductions. If an idea presents itself to us, we must not reject it simply because it does not agree with the logical deductions of a reigning theory. We may follow our feelings and our idea and give free rein to our imagination, as long as all our ideas are mere pretexts for devising new experiments that may supply us with convincing or unexpected and fertile facts.

The freedom which experimenters maintain is founded, as I said, on philosophic doubt. Indeed, we must be aware of the uncertainty of our reasonings on account of the obscurity of their starting point. The starting point, fundamentally, always rests on hypotheses or theories more or less imperfect, according to the state of development of the sciences. In biology, and especially in medicine, theories are so precarious that the experimenter maintains almost all his freedom. In chemistry and physics the facts are simpler, the sciences are more advanced, the theories more secure, and the experimenter must take more account of them and allow greater importance to the deductions of experimental reasoning based on them. But still he must never accept these theories at their face value. In our day, we have seen great physicists make discoveries of the first rank by means of experiments devised in a way that lacked all logical relation to admitted theories. Astronomers have enough confidence in the principles of their science to build up mathematical theories with them, but that does not prevent them from testing and verifying them by direct observations; this very precept, as we have seen, must not be neglected in theoretical mechanics. But in mathematics, when we start from an axiom or principle whose truth is absolutely necessary and conscious, freedom no longer exists; truths once established are immutable. Geometricians are not free to question whether the three angles of a triangle are or are not equal to two right angles; consequently they are not free to reject the logical consequences deduced from this principle.

If a doctor imagined that his reasoning had the value of a mathematician's, he would be utterly in error and would be led into the most unsound conclusions. This is unluckily what has happened and still happens to the men whom I shall call systematizers. These men start, in fact, from an idea which is based more or less on observation, and which they regard as an absolute truth. They then reason logically and without experimenting, and from deduction to deduction they succeed in building a system which is logical, but which has no sort of scientific reality. Superficial persons often let themselves be dazzled by this appearance of logic; and discussions worthy of ancient scholasticism are thus sometimes renewed in our day. The excessive faith in reasoning, which leads physiologists to a false simplification of things, comes, on the one hand, from ignorance of the science of which they speak, and, on the other hand, from lack of a feeling for the complexity of natural phenomena. That is why we sometimes see pure mathematicians, with very great minds too, fall into mistakes of this kind; they simplify too much and reason about phenomena as they construct them in their minds, but not as they exist in nature.

The great experimental principle, then, is doubt, that philosophic doubt which leaves to the mind its freedom and initiative, and from which the virtues most valuable to investigators in physiology and medicine are derived. We must trust our observations or our theories only after experimental verification. If we trust too much, the mind becomes bound and cramped by the results of its own reasoning; it no longer has freedom of action, and so lacks the power to break away from that blind faith in theories which is only scientific superstition.

It has often been said that, to make discoveries, one must be ignorant. This opinion, mistaken in itself, nevertheless conceals a truth. It means that it is better to know nothing than to keep in mind fixed ideas based on theories whose confirmation we constantly seek, neglecting meanwhile everything that fails to agree with them.

Nothing could be worse than this state of mind; it is the very opposite of inventiveness. Indeed a discovery is generally an unforeseen relation not included in theory, for otherwise it would be foreseen. In this respect, indeed, an uneducated man, knowing nothing of theory, would be in a better attitude of mind; theory would not embarrass him and would not prevent him from seeing new facts unperceived by a man preoccupied with an exclusive theory. But let us hasten to say that we certainly do not mean to raise ignorance into a principle. The better educated we are and the more acquired information we have, the better prepared shall we find our minds for making great and fruitful discoveries. Only we must keep our freedom of mind, as we said above, and must believe that in nature what is absurd, according to our theories, is not always impossible.

Men who have excessive faith in their theories or ideas are not only ill prepared for making discoveries; they also make very poor observations. Of necessity, they observe with a preconceived idea, and when they devise an experiment, they can see, in its results, only a confirmation of their theory. In this way they distort observation and often neglect very important facts because they do not further their aim. This is what made us say elsewhere that we must never make experiments to confirm our ideas, but simply to control them; [3] which means, in other terms, that one must accept the results of experiments as they come, with all their unexpectedness and irregularity.

But it happens further quite naturally that men who believe too firmly in their theories, do not believe enough in the theories of others. So the dominant idea of these despisers of their fellows is to find others' theories faulty and to try to contradict them. The difficulty, for science, is still the same. They make experiments only to destroy a theory, instead of to seek the truth. At the same time, they make poor observations, because they choose among the results of their experiments only what suits their object, neglecting whatever is unrelated to it, and carefully setting aside everything which might tend toward the idea they wish to combat. By these two opposite roads, men are thus led to the same result, that is, to falsify science and the facts.

[3] Claude Bernard, *Leçons sur les propriétés et les altérations des liquides de l'organisme.* Paris, 1859, *Première leçon.*

Accordingly, we must disregard our own opinion quite as much as the opinion of others, when faced by the decisions of experience. If men discuss and experiment, as we have just said, to prove a preconceived idea in spite of everything, they no longer have freedom of mind, and they no longer search for truth. Theirs is a narrow science, mingled with personal vanity or the diverse passions of man. Pride, however, should have nothing to do with all these vain disputes. When two physiologists or two doctors quarrel, each to maintain his own ideas or theories, in the midst of their contradictory arguments, only one thing is absolutely certain: that both theories are insufficient, and neither of them corresponds to the truth. The truly scientific spirit, then, should make us modest and kindly. We really know very little, and we are all fallible when facing the immense difficulties presented by investigation of natural phenomena. The best thing, then, for us to do is to unite our efforts, instead of dividing them and nullifying them by personal disputes. In a word, the man of science wishing to find truth must keep his mind free and calm, and if it be possible, never have his eye bedewed, as Bacon says, by human passions.

In scientific education, it is very important to differentiate, as we shall do later, between determinism which is the absolute principle of science, and theories which are only relative principles to which we should assign but temporary value in the search for truth. In a word, we must not teach theories as dogmas or articles of faith. By exaggerated belief in theories, we should give a false idea of science; we should overload and enslave the mind, by taking away its freedom, smothering its originality and infecting it with the taste for systems.

The theories which embody our scientific ideas as a whole are, of course, indispensable as representations of science. They should also serve as a basis for new ideas. But as these theories and ideas are by no means immutable truth, one must always be ready to abandon them, to alter them or to exchange them as soon as they cease to represent the truth. In a word, we must alter theory to adapt it to nature, but not nature to adapt it to theory.

To sum up, two things must be considered in experimental science: method and idea. The object of method is to direct the idea which arises in the interpretation of natural phenomena and in the search for truth. The idea must always remain independent, and we must no more chain it with scientific beliefs than with philosophic or religious beliefs; we must be bold and free in setting forth our ideas, must follow our feeling, and must on no account linger too long in childish fear of contradicting theories. If we are thoroughly steeped in the principles of the experimental method, we have nothing to fear; for, as long as the idea is correct, we go on developing it; when it is wrong, experimentation is there to set it right. We must be able, then, to attack questions even at the risk of going wrong. We do science better service, as has been said, by mistakes than by confusion, which means that we must fearlessly push ideas to their full development, provided that we regulate them and are always careful to judge them by experiment. The idea, in a word, is the motive of all reasoning, in science as elsewhere. But everywhere the idea must be submitted to a criterion. In science the criterion is the experimental method or experiment; this criterion is indispensable, and we must apply it to our own ideas as well as to those of others.

IV. The Independent Character of the Experimental Method

From all that has so far been said, it follows necessarily, that no man's opinion, formulated in a theory or otherwise, may be deemed to represent the whole truth in the sciences. It is a guide, a light, but not an absolute authority. The revolution which the experimental method has effected in the sciences is this: it has put a scientific criterion in the place of personal authority.

The experimental method is characterized by being dependent only on itself, because it includes within itself its criterion,—experience. It recognizes no authority other than that of facts and is free from personal authority. When Descartes said that we must trust only to evidence or to what is sufficiently proved, he meant that we must no longer defer to authority, as scholasticism did, but must rely only on facts firmly established by experience.

The result of this is that when we have put forward an idea or a theory in science, our object must not be to preserve it by seeking everything that may support it and setting aside everything that may weaken it. On the contrary, we ought to examine with the greatest care the facts which apparently would overthrow it, because real progress always consists in exchanging an old theory which includes fewer facts for a new one which includes more. This proves that we have advanced, for in science the best precept is to alter and exchange our ideas as fast as science moves ahead. Our ideas are only intellectual instruments which we use to break into phenomena; we must change them when they have served their purpose, as we change a blunt lancet that we have used long enough.

The ideas and theories of our predecessors must be preserved only in so far as they represent the present state of science, but they are obviously destined to change, unless we admit that science is to make no further progress, and that is impossible. In this connection, we should perhaps make a distinction between mathematical sciences and experimental sciences. As mathematical truths are immutable and absolute, the science of mathematics grows by simple successive juxtaposition of all acquired truths. As truths in the experimental sciences, on the contrary, are only relative, these sciences can move forward only by revolution and by recasting old truths in a new scientific form.

In the experimental sciences, a mistaken respect for personal authority would be superstition and would form a real obstacle to the progress of science: at the same time, it would be contrary to the examples given us by the great men of all time. Great men, indeed, are precisely those who bring with them new ideas and destroy errors. They do not, therefore, respect the authority of their own predecessors, and they do not expect us to treat them otherwise.

This non-submission to authority, which the experimental method regards as a fundamental precept, is by no means out of harmony with the respect and admiration which we bear to the great men preceding us, to whom we owe the discoveries at the base of the sciences of to-day.[4]

In the experimental sciences, great men are never the promoters of absolute and immutable truths. Each great man belongs to his time and can come only at his proper moment, in the sense that there is a necessary and ordered sequence in the appearance of scientific discoveries. Great men may be compared to torches shining at long intervals, to guide the advance of science. They light up their time, either by discovering unexpected and fertile

[4] Claude Bernard, *Cours de médecine expérimentale, leçon d'ouverture* (*Gazette méd.*, April 15, 1864.)

phenomena which open up new paths and reveal unknown horizons, or by generalizing acquired scientific facts and disclosing truths which their predecessors had not perceived. If each great man makes the science which he vitalizes take a long step forward, he never presumes to fix its final boundaries, and he is necessarily destined to be outdistanced and left behind by the progress of successive generations. Great men have been compared to giants upon whose shoulders pygmies have climbed, who nevertheless see further than they. This simply means that science makes progress subsequently to the appearance of great men, and precisely because of their influence. The result is that their successors know many more scientific facts than the great men themselves had in their day. But a great man is, none the less, still a great man, that is to say,—a giant.

There are, indeed, two sides to science in evolution: on the one hand, what is acquired already, and on the other hand, what remains to be acquired. In the already acquired, all men are more or less equal, and the great cannot be distinguished from the rest. Mediocre men often have the most acquired knowledge. It is in the darker regions of science that great men are recognized; they are marked by ideas which light up phenomena hitherto obscure and carry science forward.

To sum up, the experimental method draws from within itself an impersonal authority which dominates science. It forces this authority even on great men, instead of seeking, like the scholastics, to prove from texts that they are infallible and that they have seen, said or thought everything discovered after them. Every period has its own sum total of errors and of truths. Certain mistakes are, in a sense, inherent in their period, so that only the subsequent progress of science can reveal them. The progress of the experimental method consists in this,—that the sum of truths grows larger in proportion as the sum of error grows less. But each one of these particular truths is added to the rest to establish more general truths. In this fusion, the names of promoters of science disappear little by little, and the further science advances, the more it takes an impersonal form and detaches itself from the past. To avoid a mistake which has sometimes been committed, I hasten to add that I mean to speak here of the evolution of science only. In art and letters, personality dominates everything. There we are concerned with a spontaneous creation of the mind, that has nothing in common with the noting of natural phenomena, in which the mind must create nothing. The past keeps all its worth in the creations of art and letters; each individuality remains changeless in time and cannot be mistaken for another. A contemporary poet has characterized this sense of the personality of art and of the impersonality of science in these words,—"Art is myself; science is ourselves."

The experimental method is the scientific method which proclaims the freedom of the mind and of thought. It not only shakes off the philosophical and theological yoke; it does not even accept any personal scientific authority. This is by no means pride and boastfulness; experimenters, on the contrary, show their humility in rejecting personal authority, for they doubt their own knowledge also and submit the authority of man to the authority of experience and of the laws of nature.

Physics and chemistry, as established sciences, offer us the independence and impersonality which the experimental method demands. But medicine is still in the shades of empiricism and suffers the consequences of its backward condition. We see it still more or less mingled with religion and with the supernatural. Superstitution and the marvellous play a great part in it. Sorcerers, somnambulists, healers by virtue of some gift from Heaven, are held as the equals of physicians. Medical personality is placed above science by physicians themselves; they seek their authority in tradition, in doctrines or in medical tact. This state of affairs is the clearest of proofs that the experimental method has by no means come into its own in medicine.

The experimental method, the free thinker's method, seeks only scientific truth. Feeling, from which everything emanates, must keep its complete spontaneity and all its freedom for putting forth experimental ideas; reason also must preserve that freedom to doubt, which forces it always to submit ideas to the test of experiment. Just as, in other human actions, feeling releases an act by putting forth the idea which gives a motive to action, so in the experimental method feeling takes the initiative through the idea. Feeling alone guides the mind and constitutes the *primum movens* of science. Genius is revealed in a delicate feeling which correctly foresees the laws of natural phenomena; but this we must never forget, that correctness of feeling and fertility of idea can be established and proved only by experiment.

V. INDUCTION AND DEDUCTION IN EXPERIMENTAL REASONING

We have so far dealt with the influence of the experimental idea. Let us now consider how the method, while always forcing upon reason the dubitative form, may guide it more safely in the search for truth.

We said elsewhere that experimental reasoning is practised on observed phenomena, or observations; but it is really applied only to the ideas which the phenomena have aroused in our mind. The essence of experimental reasoning, then, will always be an idea which we introduce into a piece of experimental reasoning in order to submit it to the criterion of facts, i.e., to experiment.

There are two forms of reasoning: first, the investigating or interrogative form used by men who do not know and who wish to learn; secondly, the demonstrating or affirmative form employed by men who know or think they know, and who wish to teach others.

Philosophers seem to have differentiated these two forms of reasoning under the names of inductive reasoning and deductive reasoning. They also accept two scientific methods: the inductive method or induction, proper to the experimental physical sciences, and the deductive method or deduction, belonging more particularly to the mathematical sciences.

It follows that the one special form of experimental reasoning with which we must deal here is induction.

Induction has been defined as the process of moving from the particular to the general, while deduction is the reverse process moving from the general to the particular. I certainly shall not presume to engage in a philosophic discussion which would here be out of place and beyond my competence; only in my capacity as experimenter I shall content myself with saying that it seems to me very difficult, in practice, to justify this distinction and clearly to separate induction from deduction. If the experimenter's mind usually proceeds by starting from particular observations and going back to principles, to laws, or to general propositions, it also necessarily proceeds from the same general propositions or laws and reaches particular facts which it deduces logically from these principles. Only, when a principle is not absolutely certain, we must always make a temporary deduction requiring experimental verification. All the seeming varieties of reasoning depend merely on the nature of the subject treated and on its greater or less complexity. But in all these cases, the human mind always works in the same way, with syllogisms; it cannot behave otherwise.

Just as man goes forward, in the natural movement of his body, only by putting one foot in front of the other, so in the natural movement of his mind, man goes forward only by putting one idea in front of another. In other words, the mind, like the body, needs a primary point of support. The body's point of support is the ground which the foot feels; the mind's point of support is the known, that is, a truth or a principle of which the mind is aware. Man can learn nothing except by going from the known to the unknown; but on the other hand, as science is not infused into man at birth, and as he knows only what he learns, we seem to be in a vicious circle, where man is condemned to inability to learn anything. He would be so, in fact, if his reason did not include a feeling for relations and for determinism, which are the criteria of truth; but in no case can he gain this truth or approach it, except through reasoning and experience.

It would be incorrect to say that deduction pertains only to mathematics and induction to the other sciences exclusively. Both forms of reasoning, investigating (inductive) and demonstrating (deductive), pertain to all possible sciences, because in all the sciences there are things that we do not know and other things that we know or think we know.

When mathematicians study subjects unfamiliar to them, they use induction, like physicists, chemists or physiologists. To prove this point, I need only cite the words of a great mathematician.

Thus Euler expresses himself in a memoir entitled: *De inductione ad plenam certitudinem evehenda:*

"Notum est plerumque numerum proprietates primum per solam inductionem observatas, quas deinceps geometrae solidis demonstrationibus confirmare elaboraverunt; quo negotio in primis Fermatius summo studio et satis felici successu fuit occupatus." [5]

The principles or theories which serve as foundations for a science, whatever it may be, have not fallen from the sky; they were necessarily reached by investigation, inductive or interrogative reasoning, as we may choose to call it. It was first necessary to observe

[5] Euler, *Acta academiae scientiarum imperialis Petropolitanae, pro anno MDCCLXXX, pars posterior,* p. 38, Par. 1.

something which happened within ourselves or outside of us. From the experimental point of view there are ideas, in the sciences, which we call *a priori,* because they are a starting point for experimental reasoning (see page 27 and the following pages), but from the point of view of ideogenesis they are really *a posteriori* ideas. In a word, induction must have been the primitive, general form of reasoning; and the ideas which philosophers and men of science constantly take for *a priori* ideas are at bottom really *a posteriori* ideas.

Mathematicians and naturalists are alike when going in search of principles. Both use induction, make hypotheses, and experiment, that is to say, make attempts to verify the accuracy of their ideas. But when mathematicians and naturalists reach their principles, then they part company. Indeed, as I have already said elsewhere, the mathematician's principle is absolute, because it is not applicable to objective reality just as it is, but to relations between things considered in extremely simple conditions which the mathematician chooses and, in some sort, creates in his mind. Now, as he is thus sure that he need not introduce into his reasoning other conditions than those which he has defined, the principle remains absolute, conscious, adequate for the mind, and his logical deduction is equally certain and absolute: he no longer requires experimental verifications; logic is enough.

A naturalist is in a very different position; the general proposition which he has reached, or the principle on which he relies, is relative and provisional, because it embodies complex relations which he is never sure that he can know. Hence, his principle is uncertain, since it is unconscious and inadequate for the mind; hence, deductions, though quite logical, always remain doubtful, and so he must necessarily appeal to experiment to verify the conclusion of his deductive reasoning. The difference between mathematicians and naturalists is capital in respect to the certainty of their principles and of the conclusions to be drawn from them; but the mechanism of deductive reasoning is exactly the same for both. Both start from a proposition; only the mathematician says: Given this starting point, such and such a particular case necessarily results. The naturalist says: If this starting point is correct, such and such a particular case will follow as a consequence.

When starting from a principle, the mathematician and the naturalist, therefore, both use deduction. Both reason by making a syllogism; only, for the naturalist the conclusion of the syllogism is doubtful and requires verification, because its principle is unconscious.[6] Such experimental or dubitative reasoning is the only kind that we can use when reasoning about natural phenomena; if we wished to suppress doubt and if we dispensed with experiment, we should no longer have any criterion by which to know whether we were in the wrong or in the right, because, I repeat, the principle is unconscious, and one must therefore appeal to our senses.

From all this I should conclude that induction and deduction belong to all the sciences. I do not believe that induction and deduction are really two forms of reasoning essentially distinct. By nature man has the feeling or idea of a principle that rules particular cases. He always proceeds instinctively from a principle, acquired or invented by hypothesis; but he can never go forward in reasoning otherwise than by syllogism, that is, by proceeding from the general to the particular.

In physiology, a given organ always works through one and the same mechanism; only, when the phenomenon occurs under different conditions or in a different environment, the function takes on a different aspect; but fundamentally its character remains the same. In my opinion there is only one way of reasoning for the mind, just as there is only one way of walking for the body. But when a man goes ahead on solid flat ground, by a straight road whose whole extent he knows and sees, he advances toward his goal at an assured and rapid pace. On the contrary, when a man follows a winding road in the dark and over unknown hilly ground, he dreads precipices and goes forward cautiously, step by step. Before taking a second step, he must make sure that he has placed his first foot on a spot that is firm, then go forward in the same way verifying experimentally, moment by moment, the solidity of the ground, and always changing the direction of his advance according to what he encounters. Such is the experimenter who must never go beyond fact in his searching, lest he risk losing his way. In the two preceding examples the man goes forward over different ground and in varied surroundings, but he goes forward none the less by the same physiological method. In the same way, when an experimenter simply deduces relations from definite phenomena by means of known

[6] i.e., Not a postulate and not exclusively an affair of the mind. Translator's note.

and established principles, his reasoning develops in a secure and necessary way, while, if he finds himself in the midst of complex relations and with the support only of vague, provisional principles, the same experimenter must then go forward cautiously and must submit to experiment each one of the ideas which he successively puts forward. But, in both these cases, the mind still reasons in the same way and by the same physiological method, only it starts from a more or less binding principle.

When any sort of phenomenon strikes us in nature, we work out our idea of the cause determining it. Man in his primal ignorance imagined divinities connected with each phenomenon. To-day men of science acknowledge forces or laws: it is they that govern phenomena. An idea that comes to us at the sight of a phenomenon is called *a priori*. Now we shall later easily show that this *a priori* idea, which rises in us *à propos* of a special fact, always contains implicitly and, in some sort, without our knowledge, a *principle* to which we tend to refer the special fact, so that when we think that we are moving from a special case to a principle, i.e., making an induction, we are really making a deduction; only the experimenter guides himself by an assumed or provisional principle which he alters moment by moment, because he is searching in almost total darkness. In proportion as we gather facts, our principles become more and more general and more secure; so we gain the certainty that we deduce. But nevertheless, in the experimental sciences, our principle must always remain provisional, because we are never certain that it includes only the facts and conditions of which we are aware. In short, our deductions are always hypothetical until verified experimentally. An experimenter, therefore, can never be in the position of the mathematician, precisely because experimental reasoning, by its very nature, is always dubitative. If we wish, we can call the experimenter's dubitative reasoning induction, and the mathematician's affirmative reasoning deduction; but the distinction will then apply to the certainty or uncertainty of our starting point in reasoning, not to the way in which we reason.

VI. Doubt in Experimental Reasoning

I will summarize the preceding paragraph by saying that there seems to me to be only one form of reasoning: deduction by syllogism.

The mind, even if it wished, could not reason otherwise, and if this were the place for it, I might try to support my proposition by physiological arguments. But to find scientific truth, we, after all, have little need to know how our mind reasons; it is enough to let it reason naturally, and in that case it will always start from a principle to reach a conclusion. All we need do here is to insist on a precept which will always forearm the mind against the countless sources of error that may be met in applying the experimental method.

This general precept, one of the foundations of the experimental method, is doubt: it is expressed by saying that the conclusion of our reasoning must always remain dubitative when the starting point or the principle is not an absolute truth. We have seen that there is no absolute truth apart from mathematical principles; in all natural phenomena the principles from which we start, like the conclusions which we reach, embody only relative truths. The experimenter's stumbling block, then, consists in thinking that he knows what he does not know, and in taking for absolute, truths that are only relative. Hence, the unique and fundamental rule of scientific investigation is reduced to doubt, as great philosophers, moreover, have already proclaimed.

Experimental reasoning is precisely the reverse of scholastic reasoning. Scholasticism must always have a fixed and indubitable starting point; and, unable to find it either in outer things or in reason, it borrows it from some irrational source, such as revelation, tradition, a conventional or an arbitrary authority. The starting point once settled, scholastics or systematizers deduce logically all the consequences, even invoking as arguments observation or experience of facts when they are favorable; the one condition is that the starting point shall remain immutable and shall not vary with their experiences and observations, but on the contrary that facts shall be so interpreted as to adapt themselves to it. Experimenters, on the contrary, never accept an immutable starting point; their principle is a postulate, all of whose consequences they logically deduce, but without ever considering it absolute or beyond the reach of experiment. The chemists' elements are elements only until proof to the contrary. All the theories which serve as starting points for physicists, chemists, and with still more reason physiologists, are true only until facts are discovered which they do not include, or which contradict them. When these contradictory facts are shown to be firmly established, far from stiffening themselves against experience, like the scholastics or systematizers, experimenters, on the contrary, hasten to safeguard their starting point, to modify their theory, because they know that this is the only way to go forward and to make progress in science. Experimenters, then, always doubt even their starting point; of necessity they keep a supple and modest mind and accept contradiction, on the one condition that it be proved. Scholastics or systematizers never question their starting point, to which they seek to refer everything; they have a proud and intolerant mind and do not accept contradiction, since they do not admit that their starting point may change. Men of system are also distinguished from men of experimental science by the fact that the first impose their idea, while the second always give it just for what it is worth. Finally, another essential characteristic, which differentiates experimental reasoning from scholastic reasoning, is the fertility of the one and the sterility of the other. The scholastic who believes himself in possession of absolute certainty comes to naught; this can easily be understood, since by his absolute principle, he puts himself outside of nature, in which everything is relative. The experimenter, on the contrary, who always doubts and who does not believe that he possesses absolute certainty about anything, succeeds in mastering the phenomena that surround him and in extending his power over nature. Man can do, then, more than he knows; and true experimental science gives him power only in showing him his ignorance. Possessing absolute truth matters little to the man of science, so long as he is certain about the relations of phenomena to one another. Indeed, our mind is so limited that we can know neither the beginning nor the end of things; but we can grasp the middle, i.e., what surrounds us closely.

Systematic or scholastic reasoning is natural to inexperienced, proud minds; it is only by thorough experimental study of nature that we succeed in acquiring the experimenter's doubting mind. That takes a long time; of those who think they are following the experimental path in physiology and in medicine, many, as we shall see later, are still scholastics. As for me, I am convinced that only study of nature can give scholars a true perception of science. Philosophy, which I consider an excellent gymnastic for the mind, has systematic and scholastic tendencies in spite of itself, which would be harmful to men of science properly so-called. After all, no method can replace that study of nature which makes true men of science: without that study, all that philosophers have said and all that I myself have repeated after them in this introduction would remain inapplicable and sterile.

I do not think, therefore, as I said above, that it is very profitable for men of science to discuss definitions of induction and of deduction, nor, for that matter, the question whether we advance by one or the other of these so-called processes of mind. Baconian induction, however, has become famous and has been made the foundation of all scientific philosophy. Bacon was a great genius, and his great restoration of the sciences is sublime as an idea; we are captivated and carried along in spite of ourselves, in reading the *Novum Organum* and the *Augmentum Scientiarum.* We are fascinated by a medley of scientific gleams, clothed in the loftiest of poetic forms. Bacon felt the sterility of scholasticism; he well understood and foresaw the importance of experiment for the future of the sciences. Yet Bacon was not a man of science, and he did not understand the mechanism of the experimental method. To prove this, it would be enough to cite the hapless attempts which he made. Bacon advises us to fly from hypotheses and theories;[7] we have seen, however, that they are auxiliaries of the method, indispensable as scaffolding is necessary in building a house. Bacon, as is always the case, had extravagant admirers and detractors. Without taking one side or the other, I will say that, while recognizing Bacon's genius, I believe no more than J. de Maistre[8] that he endowed the human intellect with a new instrument, and it seems to me, as to M. de Rémusat,[9] that induction does not differ from the syllogism. Moreover, I believe that great experimenters appeared before all precepts of experimentation, as great orators preceded all treatises on rhetoric. Consequently, even in speaking of Bacon, it does not seem to me permissible to say he invented the experimental method, that method which Galileo and Torricelli so admirably practised and which Bacon never could use.

When Descartes[10] starts from universal doubt and repudiates

[7] Bacon, *Œuvres*, edition, François Riaux, Introduction, p. 30.

[8] J. de Maistre, *Examen de la philosophie de Bacon.*

[9] M. de Rémusat, *Bacon, sa vie, son temps et sa philosophie*, 1857.

[10] Descartes, *Discours sur la méthode.*

authority, he gives much more practical precepts for the experimenter than those that Bacon gives for induction. We have seen, indeed, that only doubt promotes experiment; it is doubt, finally, which determines the form of experimental reasoning.

In connection with medicine and the physiological sciences, however, it is important to determine at what point to apply doubt, so as to distinguish it from scepticism, and to show how scientific doubt becomes an element of the greatest certainty. The sceptic disbelieves in science and believes in himself; he believes enough in himself to dare deny science and to assert that it is not subject to definite, fixed laws. The doubter is a true man of science; he doubts only himself and his interpretations, but he believes in science; in the experimental sciences, he even accepts a criterion or absolute scientific principle. This principle is the determinism of phenomena, which is as absolute in the phenomena of living bodies as in those of inorganic matter, as we shall later assert (page 65).

Finally, in concluding this section, we may say that in all experimental reasoning there are two possibilities: either the experimenter's hypothesis will be disproved or it will be proved by experiment. When experiment disproves his preconceived idea, the experimenter must discard or modify it. But even when experiment fully proves his preconceived idea, the experimenter must still doubt; for since he is dealing with an unconscious truth, his reason still demands a counterproof.

VII. The Principle of the Experimental Criterion

We have just said that one must doubt, but by no means be sceptical. A sceptic, indeed, who believes nothing, no longer has a foundation on which to establish his criterion, and consequently he finds it impossible to build up a science; the sterility of his unhappy mind results at once from the error of his perception and from the imperfection of his reason. After having posited the principle that investigators must doubt, we added that doubt will apply only to the soundness of their opinions, or of their ideas as experimenters, or to the value of their means of investigation, as observers, but never to determinism, the very principle of experimental science. Let us return in a few words to this fundamental point.

Experimenters must doubt their intuition, i.e., the *a priori* idea or the theory which serves as their starting point; this is why it is an absolute principle always to submit one's idea to the experimental criterion so as to test its value. But just what is the foundation of this experimental criterion? This question may seem superfluous, after having repeatedly said that facts judge the idea and give us experience. Facts alone are real, it is said; and we must leave the matter to them, wholly and exclusively. Again, it is a fact, a sheer fact, men often repeat; there is no use in discussing, we must accept it. Of course I admit that facts are the only realities that can give form to the experimental idea and at the same time serve as its control; but this is on condition that reason accepts them. I think that blind belief in fact, which dares to silence reason, is as dangerous to the experimental sciences as the beliefs of feeling or of faith which also force silence on reason. In a word, in the experimental method as in everything else, the only real criterion is reason.

A fact is nothing in itself, it has value only through the idea connected with it or through the proof it supplies. We have said elsewhere that, when one calls a new fact a discovery, the fact itself is not the discovery, but rather the new idea derived from it; in the same way, when a fact proves anything, the fact does not itself give the proof, but only the rational relation which it establishes between the phenomenon and its cause. This relation is the scientific truth which we now must discuss further.

Let us recall how we characterized mathematical truths and experimental truths. Mathematical truths, once acquired, we said, are conscious and absolute truths, because the ideal conditions in which they exist are also conscious and known by us in an absolute way. Experimental truths, on the contrary, are unconscious and relative, because the real conditions on which they exist are unconscious and can be known by us only in their relation to the present state of our science. But if the experimental truths, which serve as foundation for our reasoning, are so wrapped up in the complex reality of natural phenomena that they appear to us only in shreds, these experimental truths rest, none the less, on principles that are absolute because, like those of mathematical truths, they speak to our consciousness and our reason. Indeed the absolute principle of experimental science is conscious and necessary determinism in the conditions of phenomena. So that, given no matter what natural phenomenon, experimenters can never acknowledge variation in the embodiment of this phenomenon, unless new conditions have at the same time occurred in its coming to pass; what is more, they have an *a priori* certainty, that these variations are determined by rigorous, mathematical relations. Experiment only shows us the form of phenomena; but the relation of a phenomenon to a definite cause is necessary and independent of experiment; it is necessarily mathematical and absolute. Thus we see that the principle of the criterion in experimental sciences is fundamentally identical with that of the mathematical sciences, since in each case the principle is expressed by a necessary and absolute relation between things. Only in the experimental sciences these relations are surrounded by numerous, complex and infinitely varied phenomena which hide them from our sight. With the help of experiment, we analyze, we dissociate these phenomena, in order to reduce them to more and more simple relations and conditions. In this way we try to lay hold on scientific truth, i.e., find the law that shall give us the key to all variations of the phenomena. Thus experimental analysis is our only means of going in search of truth in the natural sciences, and the absolute determinism of phenomena, of which we are conscious *a priori*, is the only criterion or principle which directs and supports us. In spite of our efforts, we are still very far from this absolute truth; and it is probable, especially in the biological sciences, that it will never be given us to see it in its nakedness. But this need not discourage us, for we are constantly nearing it; and moreover, with the help of our experiments, we grasp relations between phenomena which, though partial and relative, allow us more and more to extend our power over nature.

It follows from the above that, if a phenomenon, in an experiment, had such a contradictory appearance that it did not necessarily connect itself with determinate causes, then reason should reject the fact as non-scientific. We should wait or by direct experiments seek the source of error which may have slipped into the observation. Indeed, there must be error or insufficiency in the observation; for to accept a fact without a cause, that is, indeterminate in its necessary conditions, is neither more nor less than the negation of science. So that, in the presence of such a fact, men of science must never hesitate; they must believe in science and doubt their means of investigation. They will, therefore, perfect their means of observation and will make every effort to get out of the darkness; but they will never deny the absolute determinism of the phenomena; because it is precisely the recognition of determinism that characterizes true men of science.

In medicine, we are often confronted with poorly observed and indefinite facts which form actual obstacles to science, in that men always bring them up, saying: it is a fact, it must be accepted. Rational science based, as we have said, on a necessary determinism, must never repudiate an accurate and well-observed fact; but on the same principle, it ought not to encumber itself with apparent facts collected without precision, and possessing no kind of meaning, which are used as a double-edged weapon to support or disprove the most diverse opinions. In short, science rejects the indeterminate; and in medicine, when we begin to base our opinions on medical tact, on inspiration, or on more or less vague intuition about things, we are outside of science and offer an example of that fanciful medicine which may involve the greatest dangers, by surrendering the health and life of the sick to the whims of an inspired ignoramus. True science teaches us to doubt and, in ignorance, to refrain.

VIII. Proof and Counterproof

We said above that experimenters, who see their ideas confirmed by an experiment, should still doubt and require a counterproof. Indeed, proof that a given condition always precedes or accompanies a phenomenon does not warrant concluding with certainty that a given condition is the immediate cause of that phenomenon. It must still be established that, when this condition is removed, the phenomenon will no longer appear. If we limited ourselves to the proof of presence alone, we might fall into error at any moment and believe in relations of cause and effect where there was nothing but simple coincidence. As we shall later see, coincidences form one of the most dangerous stumbling blocks encountered by experimental scientists in complex sciences like biology. It is the *post hoc, ergo propter hoc* of the doctors, into which we may very easily let ourselves be led, especially if the result of an experiment or an observation supports a preconceived idea.

Counterproof, then, is a necessary and essential characteristic of the conclusion of experimental reasoning. It is the expression of philosophic doubt carried as far as possible. Counterproof decides whether the relation of cause to effect, which we seek in phenomena, has been found. To do this, it removes the accepted cause, to see if the effect persists, relying on that old and absolutely true adage: *sublata causa, tollitur effectus.* This is what we still call the *experimentum crucis.*

We must not confuse a counterexperiment or counterproof with what has been called comparative experiment. As we shall later see, this is only a comparative observation resorted to, in complex circumstances, to simplify phenomena and to forearm oneself against unforeseen sources of error; counterproof, on the contrary, is a counterjudgment dealing directly with the experimental conclusion and forming one of its necessary terms. Indeed, proof, in science, never establishes certainty without counterproof. Analysis can be absolutely proved only when the synthesis, which demonstrates it, provides the counterproof or counterexperiment. Similarly a synthesis made at the outset should be demonstrated later by analysis. Feeling for this necessary, experimental counterproof constitutes the scientific feeling *par excellence.* It is familiar to physicists and chemists; but it is far from being as well understood by physicians. In most cases, when we see two phenomena in physiology or medicine going together and following one another in a constant order, we think we may conclude that the first is the cause of the second. This would be a false judgment in very many cases; statistical tables of presence or of absence never establish experimental demonstrations. In complex sciences like medicine, we must at the same time make use of comparative experiment and of counterproof. Some physicians fear and avoid counterproof; as soon as they make observations in the direction of their ideas, they refuse to look for contradictory facts, for fear of seeing their hypothesis vanish. We have already said that this is a very poor spirit; if we mean to find truth, we can solidly settle our ideas only by trying to destroy our own conclusions by counterexperiments. Now the only proof that one phenomenon plays the part of cause in relation to another is by removing the first, to stop the second.

I shall not further emphasize this principle of the experimental method at this point, because I shall later take the opportunity to return to it, giving special examples which will explain my thought. Let me summarize by saying that experimenters should always push their investigation to the point of counterproof; without that, their experimental reasoning would not be complete. Counterproof establishes the necessary determinism of phenomena; and thus alone can satisfy reason to which, as we have said, we must always bring back any true scientific criterion.

Experimental reasoning, whose different terms we have examined in the preceding section, sets itself the same goal in all the sciences. Experimenters try to reach determinism; with the help of reasoning and of experiment they try to connect natural phenomena with their necessary conditions or, in other words, with their immediate causes. By this means, they reach the law which enables them to master phenomena. All natural philosophy is summarized in *knowing the law of phenomena*. The whole experimental problem may be reduced to foreseeing and directing phenomena. But this double goal can be attained, in living bodies, only by certain special principles of experimentation which we must point out in the following chapters.

recapitulation

computer simulation a way in physical science to experiment with different conditions

PART TWO

EXPERIMENTATION WITH LIVING BEINGS

CHAPTER I

EXPERIMENTAL CONSIDERATIONS COMMON TO LIVING THINGS AND INORGANIC BODIES

I. THE SPONTANEITY OF LIVING BEINGS IS NO OBSTACLE TO THE
USE OF EXPERIMENTATION

naturally occurring

THE spontaneity enjoyed by beings endowed with life has been one of the principal objections urged against the use of experimentation in biological studies. Every living being indeed appears to us provided with a kind of inner force, which presides over manifestations of life more and more independent of general cosmic influence in proportion as the being rises higher in the scale of organization. In the higher animals and in man, for instance, this vital force seems to result in withdrawing the living being from general physico-chemical influences and thus making the experimental approach very difficult.

Inorganic bodies offer no parallel; whatever their nature, they are all devoid of spontaneity. As the manifestation of their properties is therefore absolutely bound up in the physico-chemical conditions surrounding them and forming their environment, it follows that the experimenter can reach them and alter them at will.

On the other hand, all the phenomena of a living body are in such reciprocal harmony one with another that it seems impossible to separate any part without at once disturbing the whole organism. Especially in higher animals, their more acute sensitiveness brings with it still more notable reactions and disturbances.

Many physicians and speculative physiologists, with certain anatomists and naturalists, employ these various arguments to attack experimentation on living beings. They assume a vital force in opposition to physico-chemical forces, dominating all the phenomena

mental climate / was such that / medical sciences

of life, subjecting them to entirely separate laws, and making the organism an organized whole which the experimenter may not touch without destroying the quality of life itself. They even go so far as to say that inorganic bodies and living bodies differ radically from this point of view, so that experimentation is applicable to the former and not to the latter. Cuvier, who shares this opinion and thinks that physiology should be a science of observation and of deductive anatomy, expresses himself thus: "All parts of a living body are interrelated; they can act only in so far as they act all together; trying to separate one from the whole means transferring it to the realm of dead substances; it means entirely changing its essence."

If the above objections were well founded, we should either have to recognize that determinism is impossible in the phenomena of life, and this would be simply denying biological science; or else we should have to acknowledge that vital force must be studied by special methods, and that the science of life must rest on different principles from the science of inorganic bodies. These ideas, which were current in other times, are now gradually disappearing; but it is essential to extirpate their very last spawn, because the so-called vitalistic ideas still remaining in certain minds are really an obstacle to the progress of experimental science.

I propose, therefore, to prove that the science of vital phenomena must have the same foundations as the science of the phenomena of inorganic bodies, and that there is no difference in this respect between the principles of biological science and those of physico-chemical science. Indeed, as we have already said, the goal which the experimental method sets itself is everywhere the same; it consists in connecting natural phenomena with their necessary conditions or with their immediate causes. In biology, since these conditions are known, physiologists can guide the manifestation of vital phenomena as physicists guide the natural phenomena, the laws of which they have discovered; but in doing so, experimenters do not act on life.

Yet there is absolute determinism in all the sciences, because every phenomenon being necessarily linked with physico-chemical conditions, men of science can alter them to master the phenomenon, i.e., to prevent or to promote its appearing. As to this, there is absolutely no question in the case of inorganic bodies. I mean to prove that it is the same with living bodies, and that for them also determinism exists.

II. Manifestation of Properties of Living Bodies Is Connected with the Existence of Certain Physico-Chemical
Phenomena Which Regulate Their Appearance

The manifestation of properties of inorganic bodies is connected
with surrounding conditions of temperature and moisture by means
of which the experimenter can directly govern mineral phenomena.
Living bodies at first sight do not seem capable of being thus
influenced by neighboring physico-chemical conditions; but that is
merely a delusion depending on the animal having and maintaining
within himself the conditions of warmth and moisture necessary to
the appearance of vital phenomena. The result is that an inert body,
obedient to cosmic conditions, is linked with all their variations,
while a living body on the contrary remains independent and free
in its manifestations; it seems animated by an inner force that rules
all its acts and liberates it from the influence of surrounding physico-
chemical variations and disturbances. This quite different aspect
of the manifestations of living bodies as compared with the behavior
of inorganic bodies has led the physiologists, called vitalists, to attribute to the former a vital force ceaselessly at war with physico-
chemical forces and neutralizing their destructive action on the living
organism. According to this view, the manifestations of life are
determined by spontaneous action of this special vital force, instead of
being, like the manifestations of inorganic bodies, the necessary results of conditions or of the physico-chemical influences of a surrounding environment. But if we consider it, we shall soon see that the
spontaneity of living bodies is simply an appearance and the
result of a certain mechanism in completely determined environments; so that it will be easy, after all, to prove that the behavior
of living bodies, as well as the behavior of inorganic bodies, is dominated by a necessary determinism linking them with conditions of a
purely physico-chemical order.

Let us note, first of all, that this kind of independence of living
beings in the cosmic environment appears only in complex higher
animals. Inferior beings, such as the infusoria, reduced to an elementary organism, have no real independence. These creatures
exhibit the vital properties with which they are endowed, only under
the influence of external moisture, light or warmth, and as soon
as one or more of these conditions happens to fail, the vital manifes-

tation ceases, because the parallel physico-chemical phenomenon has stopped. In vegetables the manifestation of vital phenomena is linked in the same way with conditions of warmth, moisture and light in the surrounding environment. It is the same again with cold-blooded animals; the phenomena of life are benumbed or stimulated according to the same conditions. Now the influences producing or retarding vital manifestations in living beings are exactly the same as those which produce, accelerate or retard manifestations of physico-chemical phenomena in inorganic bodies, so that instead of following the example of the vitalists in seeing a kind of opposition or incompatibility between the conditions of vital manifestations and the conditions of physico-chemical manifestations, we must note, on the contrary, in these two orders of phenomena a complete parallelism and a direct and necessary relation. Only in warm-blooded animals do the conditions of the organism and those of the surrounding environment seem to be independent; in these animals indeed the manifestation of vital phenomena no longer suffers the alternations and variations that the cosmic conditions display; and an inner force seems to join combat with these influences and in spite of them to maintain the vital forces in equilibrium. But fundamentally it is nothing of the sort; and the semblance depends simply on the fact that, by the more complete protective mechanism which we shall have occasion to study, the warm-blooded animal's internal environment comes less easily into equilibrium with the external cosmic environment. External influences, therefore, bring about changes and disturbances in the intensity of organic functions only in so far as the protective system of the organism's internal environment becomes insufficient in given conditions.

III. Physiological Phenomena in the Higher Animals Take Place in Perfected Internal Organic Environments Endowed with Constant Physico-Chemical Properties

Thoroughly to understand the application of experimentation to living beings, it is of the first importance to reach a definite judgment on the ideas which we are now explaining. When we examine a higher, i.e., a complex living organism, and see it fulfill its different functions in the general cosmic environment common to all the phenomena of nature, it seems to a certain extent independent of this environment. But this appearance results simply from our deluding ourselves about the simplicity of vital phenomena. [The external phenomena which we perceive in the living being are fundamentally very complex; they are the resultant of a host of intimate properties of organic units whose manifestations are linked together with the physico-chemical conditions of the internal environment in which they are immersed. [In our explanations we suppress this inner environment and see only the outer environment before our eyes. But the real explanation of vital phenomena rests on study and knowledge of the extremely tenuous and delicate particles which form the organic units of the body.] This idea, long ago set forth in biology by great physiologists, seems more and more true in proportion as the science of the organization of living beings makes progress. We must, moreover, learn that the *intimate particles* of an organism exhibit their vital activity only through a necessary physico-chemical relation with immediate environments which we must also study and know. Otherwise, if we limit ourselves to the survey of total phenomena visible from without, we may falsely believe that a force in living beings violates the physico-chemical laws of the general cosmic environment, just as an untaught man might believe that some special force in a machine, rising in the air or running along the ground, violated the laws of gravitation. Now, a living organism is nothing but a wonderful machine endowed with the most marvellous properties and set going by means of the most complex and delicate mechanism. There are no forces opposed and struggling one with another; in nature there can be only order and disorder, harmony or discord.

[In experimentation on inorganic bodies, we need take account of only one environment, the external cosmic environment; while in the higher living animals, at least two environments must be considered, the external or extra-organic environment and the internal or intra-organic environment.] In my course on physiology at the Faculty of Sciences, I explain each year these ideas on organic environment,— new ideas which I regard as fundamental in general physiology; they are also necessarily fundamental in general pathology, and the same thoughts will guide us in adapting experimentation to living beings, For, as I have said elsewhere, the great difficulties that we meet in experimentally determining vital phenomena and in applying suitable means to altering them are caused by the complexity involved in the existence of an internal organic environment.[1]

Physicists and chemists experimenting on inert bodies need consider only the external environment; by means of the thermometer, barometer and other instruments used in recording and measuring the properties of the external environment, they can always set themselves in equivalent conditions. For physiologists these instruments no longer suffice; and yet the internal environment is just the place where they should use them. Indeed, the internal environment of living beings is always in direct relation with the normal or pathological vital manifestations of organic units. In proportion as we ascend the scale of living beings, the organism grows more complex, the organic units become more delicate and require a more perfected internal environment. The circulating liquids, the blood serum and the intra-organic fluids all constitute the internal environment.

In living beings the internal environment, which is a true product of the organism, preserves the necessary relations of exchange and equilibrium with the external cosmic environment; but in proportion as the organism grows more perfect, the organic environment becomes specialized and more and more isolated, as it were, from the surrounding environment. In vegetables and in cold-blooded animals, as we have said, this isolation is less complete than in warm-blooded animals; in the latter the blood serum maintains an almost fixed and constant temperature and composition. But these differing conditions do not constitute differences of nature in different living beings; they are merely improvements in the isolating and protecting mechanisms of their environment. Vital manifestations in animals vary only because the physico-chemical conditions of their internal environments vary; thus a mammal, whose blood has been chilled either by natural hibernation or by certain lesions of the nervous system, closely resembles a really cold-blooded animal in the properties of its tissues.

To sum up, from what has been said we can gain an idea of the enormous complexity of vital phenomena and of the almost insuperable difficulties which their accurate determination opposes to physiologists forced to carry on experimentation in the internal or organic

---

[1] Claude Bernard, *Leçons sur la physiologie et la pathologie du système nerveux. Leçon d'ouverture*, Dec. 17, 1856. Paris, 1858, Vol. I.—*Cours de pathologie expérimentale.* (*The Medical Times*, 1860.)

environments. These obstacles, however, cannot terrify us if we are thoroughly convinced that we are on the right road. Absolute determinism exists indeed in every vital phenomenon; hence biological science exists also; and consequently the studies to which we are devoting ourselves will not all be useless. General physiology is the basic biological science toward which all others converge. Its problem is to determine the elementary condition of vital phenomena. Pathology and therapeutics also rest on this common foundation. By normal activity of its organic units, life exhibits a state of health; by abnormal manifestation of the same units, diseases are characterized; and finally through the organic environment modified by means of certain toxic or medicinal substances, therapeutics enables us to act on the organic units. To succeed in solving these various problems, we must, as it were, analyze the organism, as we take apart a machine to review and study all its works; that is to say, before succeeding in experimenting on smaller units we must first experiment on the machinery and on the organs. We must, therefore, have recourse to analytic study of the successive phenomena of life, and must make use of the same experimental method which physicists and chemists employ in analyzing the phenomena of inorganic bodies. The difficulties which result from the complexity of the phenomena of living bodies arise solely in applying experimentation; for fundamentally the object and principles of the method are always exactly the same.

IV. The Aim of Experimentation Is the Same in Study of Phenomena of Living Bodies as in Study of Phenomena of Inorganic Bodies

If the physicist and the physiologist differ in this, that one busies himself with phenomena taking place in inorganic matter, and the other with phenomena occurring in living matter, still they do not differ in the object which they mean to attain. Indeed, they both set themselves a common object, viz., getting back to the immediate cause of the phenomena which they are studying.

Now, what we call the immediate cause of a phenomenon is nothing but the physical and material condition in which it exists or appears. The object of the experimental method or the limit of every scientific research is therefore the same for living bodies as for inorganic bodies; it consists in finding the relations which connect any phenomenon with its immediate cause, or putting it differently, it consists in defining the conditions necessary to the appearance of the phenomenon. Indeed, when an experimenter succeeds in learning the necessary conditions of a phenomenon, he is, in some sense, its master; he can predict its course and its appearance, he can promote or prevent it at will. An experimenter's object, then, is reached; through science, he has extended his power over a natural phenomenon.

We shall therefore define physiology thus: the science whose object it is to study the phenomena of living beings and to *determine* the material conditions in which they appear. Only by the analytic or experimental method can we attain the determination of the conditions of phenomena, in living bodies as well as in inorganic bodies; for we reason in identically the same way in experimenting in all the sciences.

For physiological experimenters, neither spiritualism nor materialism can exist. These words belong to a philosophy which has grown old; they will fall into disuse through the progress of science. We shall never know either spirit or matter; and if this were the proper place I should easily show that on one side, as on the other, we quickly fall into scientific negations. The conclusion is that all such considerations are idle and useless. It is our sole concern to study phenomena, to learn their material conditions and manifestations, and to determine the laws of those manifestations.

First causes are outside the realm of science; they forever escape us in the sciences of living as well as in those of inorganic bodies. The experimental method necessarily turns aside from the chimerical search for a vital principle; vital force exists no more than mineral force exists, or, if you like, one exists quite as much as the other. The word, force, is merely an abstraction which we use for linguistic convenience. For mechanics, force is the relation of a movement to its cause. For physicists, chemists and physiologists, it is fundamentally the same. As the essence of things must always remain unknown, we can learn only relations, and phenomena are merely the results of relations. The properties of living bodies are revealed only through reciprocal organic relations. A salivary gland, for instance, exists only because it is in relation with the digestive system, and because its histological units are in certain relations one with another and with the blood. Destroy these relations by isolating the units of the organism, one from another in thought, and the salivary gland simply ceases to be.]

A scientific law gives us the numerical relation of an effect to its cause, and that is the goal at which science stops. When we have the law of a phenomenon, we not only know absolutely the conditions determining its existence, but we also have the relations applying to all its variations, so that we can predict modifications of the phenomenon in any given circumstances.

[As a corollary to the above we must add that neither physiologists nor physicians need imagine it their task to seek the cause of life or the essence of disease. That would be entirely wasting one's time in pursuing a phantom. The words, life, death, health, disease, have no objective reality. We must imitate the physicists in this matter and say, as Newton said of gravitation: "Bodies fall with an accelerated motion whose law we know: that is a fact, that is reality. But the first cause which makes these bodies fall is utterly unknown. To picture the phenomenon to our minds, we may say that the bodies fall as if there were a force of attraction toward the centre of the earth, *quasi esset attractio*. But the force of attraction does not exist, we do not see it; it is merely a word used to abbreviate speech." When a physiologist calls in vital force or life, he does not see it; he merely pronounces a word; only the vital phenomenon exists, with its material conditions; that is the one thing that he can study and know.

To sum up, the object of science is everywhere the same: to learn the material conditions of phenomena. But though this goal is the same in the physico-chemical and in biological sciences, it is much harder to reach in the latter because of the mobility and complexity of the phenomena which we meet.

V. The Necessary Conditions of Natural Phenomena Are Absolutely Determined in Living Bodies as Well as in Inorganic Bodies

We must acknowledge as an experimental axiom that in living beings as well as in inorganic bodies the necessary conditions of every phenomenon are absolutely determined. That is to say, in other terms, that when once the conditions of a phenomenon are known and fulfilled, the phenomenon must always and necessarily be reproduced at the will of the experimenter. Negation of this proposition would be nothing less than negation of science itself. Indeed, as science is simply the determinate and the determinable, we must perforce accept as an axiom that, in identical conditions, all phenomena are identical and that, as soon as conditions are no longer the same, the phenomena cease to be identical. This principle is absolute in the phenomena of inorganic bodies as well as in those of living beings, and the influence of life, whatever view of it we take, can nowise alter it. As we have said, what we call vital force is a first cause analogous to all other first causes, in this sense, that it is utterly unknown. It matters little whether or not we admit that this force differs essentially from the forces presiding over manifestations of the phenomena of inorganic bodies, the vital phenomena which it governs must still be determinable; for the force would otherwise be blind and lawless, and that is impossible. The conclusion is that the phenomena of life have their special law because there is rigorous determinism in the various circumstances constituting conditions necessary to their existence or to their manifestations; and that is the same thing. Now in the phenomena of living bodies as in those of inorganic bodies, it is only through experimentation, as I have already often repeated, that we can attain knowledge of the conditions which govern these phenomena and so enable us to master them.

Everything so far said may seem elementary to men cultivating the physico-chemical sciences. But among naturalists and especially among physicians, we find men who, in the name of what they call vitalism, express most erroneous ideas on the subject which concerns us. They believe that study of the phenomena of living matter can have no relation to study of the phenomena of inorganic matter. They look on life as a mysterious supernatural influence which acts arbitrarily by freeing itself wholly from determinism, and they brand as materialists all who attempt to reconcile vital phenomena with definite organic and physico-chemical conditions. These false ideas are not easy to uproot when once established in the mind; only the progress of science can dispel them. But vitalistic ideas, taken in the sense which we have just indicated, are just a kind of medical superstition,—a belief in the supernatural. Now, in medicine, belief in occult causes, whether it is called vitalism or is otherwise named, encourages ignorance and gives birth to a sort of unintentional quackery; that is to say, the belief in an inborn, indefinable science. Con-

if determinism is not found you must find by continued investigation

fidence in absolute determinism in the phenomena of life leads, on the contrary, to real science, and gives the modesty which comes from the consciousness of our little learning and the difficulty of science. This feeling incites us, in turn, to work toward knowledge; and to this feeling alone, science in the end owes all its progress.

I should agree with the vitalists if they would simply recognize that living beings exhibit phenomena peculiar to themselves and unknown in inorganic nature. I admit, indeed, that manifestations of life cannot be wholly elucidated by the physico-chemical phenomena known in inorganic nature. I shall later explain my view of the part played in biology by physico-chemical sciences; I will here simply say that if vital phenomena differ from those of inorganic bodies in complexity and appearance, this difference obtains only by virtue of determined or determinable conditions proper to themselves. So if the sciences of life must differ from all others in explanation and in special laws, they are not set apart by scientific method. Biology must borrow the experimental method of physico-chemical sciences, but keep its special phenomena and its own laws.

In living bodies, as in inorganic bodies, laws are immutable, and the phenomena governed by these laws are bound to the conditions on which they exist, by a necessary and absolute determinism. I use the word determinism here as more appropriate than the word fatalism, which sometimes serves to express the same idea. Determinism in the conditions of vital phenomena should be one of the axioms of experimenting physicians. If they are thoroughly imbued with the truth of this principle, they will exclude all supernatural intervention from their explanations; they will have unshaken faith in the idea that fixed laws govern biological science; and at the same time they will have a reliable criterion for judging the often variable and contradictory appearance of vital phenomena. Indeed, starting with the principle that immutable laws exist, experimenters will be convinced that phenomena can never be mutually contradictory, if they are observed in the same conditions; and if they show variations, they will know that this is necessarily so because of the intervention or interference of other conditions which alter or mask phenomena. There will be occasion thenceforth to try to learn the conditions of these variations, for there can be no effect without a cause. Determinism thus becomes the foundation of all scientific progress and criticism. If we find disconcerting or even contradictory results in performing

no exception
it must be reproducible

an experiment, we must never acknowledge exceptions or contradictions as real. That would be unscientific. We must simply and necessarily decide that conditions in the phenomena are different, whether or not we can explain them at the time.

I assert that the word exception is unscientific; and as soon as laws are known, no exception indeed can exist, and this expression, like so many others, merely enables us to speak of things whose causation we do not know. Every day we hear physicians use the words: ordinarily, more often, generally, or else express themselves numerically by saying, for instance: nine times out of ten, things happen in this way. I have heard old practitioners say that the words "always" and "never" should be crossed out of medicine. I condemn neither these restrictions nor the use of these locutions if they are used as empirical approximations about the appearances of phenomena when we are still more or less ignorant of the exact conditions in which they exist. But certain physicians seem to reason as if exceptions were necessary; they seem to believe that a vital force exists which can arbitrarily prevent things from always happening alike; so that exceptions would result directly from the action of mysterious vital force. Now this cannot be the case; what we now call an exception is a phenomenon, one or more of whose conditions are unknown; if the conditions of the phenomena of which we speak were known and determined, there would be no further exceptions, medicine would be as free from them as is any other science. For instance, we might formerly say that sometimes the itch was cured and sometimes not; but now that we attack the cause of this disease, we cure it always. Formerly it might be said that a lesion of the nerves brought on paralysis, now of feeling, and again of motion; but now we know that cutting the anterior spinal nerve paralyzes motion only. Motor paralysis occurs consistently and always, because its condition has been accurately determined by experimenters.

The certainty with which phenomena are determined should also be, as we have said, the foundation of experimental criticism, whether applied to one's self or to others. A phenomenon, indeed, always appears in the same way if conditions are similar; the phenomenon never fails if the conditions are present, just as it does fail to appear if the conditions are absent. Thus an experimenter who has made an experiment, in conditions which he believes were determined, may happen not to get the same results in a new series of investigations

determinism - the process of setting up and solving problems.

as in his first observation; in repeating the experiment, with fresh precautions, it may happen again that, instead of his first result, he may encounter a wholly different one. In such a situation, what is to be done? Should we acknowledge that the facts are indeterminable? Certainly not, since that cannot be. We must simply acknowledge that experimental conditions, which we believed to be known, are not known. We must more closely study, search out and define the experimental conditions, for the facts cannot be contradictory one to another; they can only be indeterminate. Facts never exclude one another, they are simply explained by differences in the conditions in which they are born. So an experimenter can never deny a fact that he has seen and observed, merely because he cannot rediscover it. In the third part of this introduction, we shall cite instances in which the principles of experimental criticism which we have just suggested, are put in practice.

VI. To Have Determinism for Phenomena, in Biological as in
Physico-Chemical Sciences, We Must Reduce the
Phenomena to Experimental Conditions as
Definite and Simple as Possible

As a natural phenomenon is only the expression of ratios and relations and connections, at least two bodies are necessary to its appearance. So we must always consider, first, a body which reacts or which manifests the phenomenon; second, another body which acts and plays the part of environment in relation to the first. It is impossible to imagine a body wholly isolated in nature; it would no longer be real, because there would be no relation to manifest its existence.

In phenomenal relations, as nature presents them to us, more or less complexity always prevails. In this respect mineral phenomena are much less complex than vital phenomena; this is why the sciences dealing with inorganic bodies have succeeded in establishing themselves more quickly. In living bodies, the complexity of phenomena is immense, and what is more, the mobility accompanying vital characteristics makes them much harder to grasp and to define.

The properties of living matter can be learned only through their relation to the properties of inorganic matter; it follows that the biological sciences must have as their necessary foundation the physico-chemical sciences from which they borrow their means of analysis and their methods of investigation. Such are the necessary reasons for the secondary and backward evolution of the sciences concerned with the phenomena of life. But though the complexity of vital phenomena creates great obstacles, we must not be appalled, for, as we have already said, unless we deny the possibility of biological science, the principles of science are everywhere the same. So we may be sure that we are on the right road and that in time we shall reach the scientific result that we are seeking, that is to say, determinism in the phenomena of living beings.

We can reach knowledge of definite elementary conditions of phenomena only by one road, viz., by experimental analysis. Analysis dissociates all the complex phenomena successively into more and more simple phenomena, until they are reduced, if possible, to just two elementary conditions. Experimental science, in fact, considers in a phenomenon only the definite conditions necessary to produce it. Physicists try to picture these conditions to themselves, more or less ideally in mechanics or mathematical physics. Chemists successively analyze complex matters; and in thus reaching either elements or definite substances (individual compounds or chemical species), they attain the elementary or irreducible conditions of phenomena. In the same way, biologists should analyze complex organisms and reduce the phenomena of life to conditions that cannot be analyzed in the present state of science.

Experimental physiology and medicine have no other goal. When faced by complex questions, physiologists and physicians, as well as physicists and chemists, should divide the total problem into simpler and simpler and more and more clearly defined partial problems. They will thus reduce phenomena to their simplest possible material conditions and make application of the experimental method easier and more certain. All the analytic sciences divide problems, in order to experiment better. By following this path, physicists and chemists have succeeded in reducing what seemed the most complex phenomena to simple properties connected with well-defined mineral species. By following the same analytic path, physiologists should succeed in reducing all the vital manifestations of a complex organism to the play of certain organs, and the action of these organs to the properties of well-defined tissues or organic units. Anatomico-physiological experimental analysis, which dates from Galen, has just this meaning, and histology, in pursuing the same problem to-day, is naturally coming closer and closer to the goal.

Though we can succeed in separating living tissues into chemical elements or bodies, still these elementary chemical bodies are not elements for physiologists. In this respect biologists are more like physicists than chemists, for they seek to determine the properties of bodies and are much less preoccupied with their elementary composition. In the present state of the science, it would be impossible to establish any relation between the vital properties of bodies and their chemical composition; because tissues and organs endowed with the most diverse properties are at times indistinguishable from the point of view of their elementary chemical composition. Chemistry is most useful to physiologists in giving them means of separating and studying individual compounds, true organic products which play important parts in the phenomena of life.

Organic individual compounds, though well defined in their properties, are still not active elements in physiological phenomena; like mineral matter, they are, as it were, only passive elements in the organism. For physiologists, the truly active elements are what we call anatomical or histological units. Like the organic individual compounds, these are not chemically simple; but physiologically considered, they are as simplified as possible in that their vital properties are the simplest that we know,—vital properties which vanish when we happen to destroy this elementary organized part. However, all ideas of ours about these elements are limited by the present state of our knowledge; for there can be no question that these histological units, in the condition of cells and fibres, are still complex. That is why certain naturalists refuse to give them the names of elements and propose to call them elementary organisms. This appellation is in fact more appropriate; we can perfectly well picture to ourselves a complex organism made up of a quantity of distinct elementary organisms, uniting, joining and grouping together in various ways, to give birth first to the different tissues of the body, then to its various organs; anatomical mechanisms are themselves only assemblages of organs which present endlessly varied combinations in living beings. When we come to analyze the complex manifestations of any organism, we should therefore separate the complex phenomena and reduce them to a certain number of simple properties belonging to elementary organisms; then synthetically reconstruct the total organism in thought, by reuniting and ordering the elementary organisms, considered at first separately, then in their reciprocal relations.

When physicians, chemists or physiologists, by successive experimental analyses, succeed in determining the irreducible element of a phenomenon in the present state of their science, the scientific problem is simplified, but its nature is not changed thereby; and men of science are no nearer to absolute knowledge of the essence of things. Nevertheless, they have gained what it is truly important to obtain, to wit, knowledge of the necessary conditions of the phenomenon and determination of the definite relation existing between a body manifesting its properties and the immediate cause of this manifestation. The object of analysis, in biological as in physico-chemical science, is, after all, to determine and, as far as possible, to isolate the conditions governing the occurrence of each phenomenon. We can act on the phenomena of nature only by reproducing the natural conditions in which they exist; and we act the more easily on these conditions in proportion as they have first been better analyzed and reduced to a greater state of simplicity. Real science exists, then, only from the moment when a phenomenon is accurately defined as to its nature and rigorously determined in relation to its material conditions, that is, when its law is known. Before that, we have only groping and empiricism.

VII. In Living Bodies, Just as in Inorganic Bodies, the Existence of Phenomena Is Always Doubly Conditioned

The most superficial examination of what goes on around us shows that all natural phenomena result from the reaction of bodies one against another. There always come under consideration the *body,* in which the phenomenon takes place, and the outward circumstance or the environment which determines or invites the body to exhibit its properties. The conjunction of these conditions is essential to the appearance of the phenomenon. If we suppress the environment, the phenomenon disappears, just as if the body had been taken away. The phenomena of life, as well as those of inorganic bodies, are thus doubly conditioned. On the one hand, we have the organism in which vital phenomena come to pass; on the other hand, the cosmic environment in which living bodies, like inorganic bodies, find the conditions essential to the appearance of their phenomena. The conditions necessary to life are found neither in the organism nor in the outer environment, but in both at once. Indeed, if we suppress or disturb the organism, life ceases, even though the environment remains intact; if, on the other hand, we take away or vitiate the environment, life just as completely disappears, even though the organism has not been destroyed.

Thus phenomena appear as results of contact or relation of a body with its environment. Indeed, if we absolutely isolate a body in our thought, we annihilate it in so doing; and if, on the contrary, we multiply its relations with the outer world, we multiply its properties.

Phenomena, then, are definite relations of bodies; we always conceive these relations as resulting from forces outside of matter, because we cannot absolutely localize them in a single body. For physicists, universal attraction is only an abstract idea; manifestation of this force requires the presence of two bodies; if only one body is present, we can no longer conceive of attraction. For example, electricity results from the action of copper and zinc in certain chemical conditions; but if we suppress the interrelation of bodies, electricity,—an abstraction without existence in itself,— ceases to appear. In the same way, life results from contact of the organism with its environment; we can no more understand it through the organism alone than through the environment alone. It is therefore a similar abstraction, that is to say, a force which appears as if it were outside of matter.

But however the mind conceives the forces of nature, that cannot alter an experimenter's conduct in any respect. For him the problem reduces itself solely to determining the material conditions in which a phenomenon appears. These conditions once known, he can then master the phenomenon; by supplying or not supplying them, he can make the phenomenon appear or disappear at will. Thus physicists and chemists exert their power over inorganic bodies; thus physiologists gain empire over vital phenomena. Living bodies, however, seem at first sight to elude the experimenter's action. We see the higher organisms uniformly exhibit their vital phenomena, in spite of variations in the surrounding cosmic environment, and from another angle we see life extinguished in an organism after a certain length of time without being able to find reasons in the external environment for this extinction. But, as we have already said, there is an illusion here, resulting from incomplete and superficial analysis of the conditions of vital phenomena. Ancient science was able to conceive only the outer environment; but to establish the science of experimental biology, we must also conceive an inner environment. I believe I was the first to express this idea clearly and to insist on it, the better to explain the application of experimentation to living beings. Since the outer environment, on the other hand, infiltrates into the inner environment, knowing the latter teaches us the former's every influence. Only by passing into the inner, can the influence of the outer environment reach us, whence it follows that knowing the outer environment cannot teach us the actions born in, and proper to, the inner environment. The general cosmic environment is common to living and to inorganic bodies; but the inner environment created by an organism is special to each living being. Now, here is the true physiological environment; this it is which physiologists and physicians should study and know, for by its means they can act on the histological units which are the only effective agents in vital phenomena. Nevertheless, though so deeply seated, these units are in communication with the outer world; they still live in the conditions of the outer environment perfected and regulated by the play of the organism. The organism is merely a living machine so constructed that, on the one hand, the outer environment is in free communication with the inner organic environment, and, on the other hand, the organic units have protective functions, to place in reserve the materials of life and uninterruptedly to maintain the humidity, warmth and other conditions essential to vital activity. Sickness and death are merely a dislocation or disturbance of the mechanism which regulates the contact of vital stimulants with organic units. In a word, vital phenomena are the result of contact between the organic units of the body with the *inner physiological environment;* this is the pivot of all experimental medicine. Physiologists and physicians gain mastery over the phenomena of life by learning which conditions, in this inner environment, are normal and which abnormal, for the appearance of vital activity in the organic units; for apart from complexity of conditions, phenomena exhibiting life, like physico-chemical phenomena, result from contact between an active body and the environment in which it acts.

VIII. In Biological as in Physico-Chemical Science, Determinism Is Possible, Because Matter in Living as in Inorganic Bodies Can Possess No Spontaneity

To sum up, the study of life includes two things: (1) Study of the properties of organized units; (2) study of the organic environment, i.e., study of the conditions which this environment must fulfill to permit the appearance of vital activities. Physiology, pathology and therapeutics rest on this double knowledge; apart from this, neither medical science nor any truly scientific or effectual therapeutics exists.

In living organisms it is convenient to distinguish between three kinds of definite bodies: first, chemical elements; second, organic and inorganic individual compounds; third, organized anatomical units. Of about 70 elements known to chemistry to-day, only 16 are found in that most complex of organisms, the organism of man. But these 16 elements combine with one another to form the various liquid, solid and gaseous substances of the organism. Oxygen and nitrogen, however, are merely dissolved in the organic fluids; and in living beings, seem to act as elements. The inorganic individual compounds (earthy salts, phosphates, chlorides, sulphates, etc.) are essential constituents in the composition of living bodies, but are taken ready-made directly from the outer world. Organic individual compounds are also constituents of living bodies, but by no means borrowed from the outer world; they are made by the vegetable or animal organism; among such substances are starch, sugar, fat, albumen, etc., etc. When extracted from the body, they preserve their properties because they are not alive; they are organic products, but not organized. Anatomical units stand alone as organized living parts. These parts are irritable and, under the influence of various stimulants, exhibit properties exclusively characteristic of living beings. They live and nourish themselves, and their nourishment creates and preserves their properties, which means that they cannot be cut off from the organism without more or less rapidly losing their vitality.

Though very different from one another in respect to their functions in the organism, these three classes of bodies all show physico-chemical reactions under the influence of the outer stimuli,—warmth, light, electricity; but living parts also have the power of being irritable, i.e., reacting under the influence of certain stimuli in a way specially characteristic of living tissues, such as muscular contraction, nervous transmission, glandular secretion, etc. But whatever the variety presented by the three classes of phenomena, whether the reaction be physico-chemical or vital, it is never in any way spontaneous. The phenomenon always results from the influence exerted on the reacting body by a physico-chemical stimulant outside itself.

Every definite substance, whether inorganic, organic or organized, is autonomous; that is to say, it has characteristic properties and exhibits independent action. Nevertheless, each one of these bodies is inert, that is, it is incapable of putting itself into action; to do this, it must always enter into relation with another body, from which it receives a stimulus. Thus every mineral body in the cosmic environment is stable; it changes its state only when the circumstances in which it is placed are rather seriously changed, either naturally or through experimental interference. In any organic environment, the substances created by animals and vegetables are much more changeable and less stable, but still they are inert and exhibit their properties only as they are influenced by agents outside themselves. Finally, anatomical units themselves, which are the most changeable and unstable of substances, are still inert, that is, they never break into vital activity unless some foreign influence invites them. A muscle-fibre, for instance, has the vital property peculiar to itself of contracting, but this living fibre is inert in the sense that if nothing changes in its environmental or its inner conditions, it cannot bring its functions into play, and it will not contract. For the muscular fibre to contract, a change must necessarily be produced in it, by its coming into relation with a stimulation from without, which may come either from the blood or from a nerve. We may say as much of all the histological units, nerve units, blood units, etc. Different living units thus play the part of stimuli, one in relation to another; and the functional manifestations of an organism are merely the expression of their harmonious reciprocal relations. The histological units react either separately or one against another by means of vital properties which are themselves in necessary connection with surrounding physico-chemical conditions; and this relation is so intimate that we may say the intensity of physico-chemical phenomena taking place in an organism may be used to measure the intensity of its vital phenomena. Therefore, as has already been said, we must not set up an antagonism between vital phenomena and physico-chemical phenomena, but, on the contrary, we must note the complete and necessary parallelism between the two classes of phenomena. To sum up, living matter is no more able than inorganic matter to get into activity or movement by itself. Every change in matter implies intervention of a new relation, i.e., an outside condition or influence. The rôle of men of science is to try to define and determine the material conditions producing the appearance of each phenomenon. These conditions once known, experimenters master the phenomenon in this sense, that they can give movement to matter, or take it away, at pleasure.

What we have just said is equally true for the phenomena of living bodies and the phenomena of inorganic bodies. Only in the case of the complex higher organisms, physiologists and physicians must study the stimuli of vital phenomena, not in the relations of the whole organism with the general cosmic environment, but rather in the organic conditions of the inner environment. Considered in the general cosmic environment, the functions of man and of the higher animals seem to us, indeed, free and independent of the physico-chemical conditions of the environment, because its actual stimuli are found in an inner, organic, liquid environment. What we see from the outside is merely the result of physico-chemical stimuli from the inner environment; that is where physiologists must build up the real determinism of vital functions.

Living machines are therefore created and constructed in such a way that, in perfecting themselves, they become freer and freer in the general cosmic environment. But the most absolute determinism still obtains, none the less, in the inner environment which is separated more and more from the outer cosmic environment, by reason of the same organic development. A living machine keeps up its movement because the inner mechanism of the organism, by acts and forces ceaselessly renewed, repairs the losses involved in the exercise of its functions. Machines created by the intelligence of man, though infinitely coarser, are built in just this fashion. A steam engine's activity is independent of outer physico-chemical conditions, since the machine goes on working through cold, heat, dryness and moisture. But physicists going down into the inner environment of the machine, find that this independence is only apparent, and that the movement of its every inner gear is determined by physical conditions whose law they know. As for physiologists, if they can go down into the inner environment of a living machine, they find likewise absolute determinism that must become the real foundation of the science of living bodies.

IX. The Limits of Our Knowledge Are the Same in the Phenomena of Living Bodies and in the Phenomena of Inorganic Bodies

The nature of our mind leads us to seek the essence or the *why* of things. Thus we aim beyond the goal that it is given us to reach; for experience soon teaches us that we cannot get beyond the *how*, i.e., beyond the immediate cause or the necessary conditions of phenomena. In this respect the limits of our knowledge are the same in biological as in physico-chemical sciences.

When, by successive analyses, we find the immediate cause determining the circumstances in which a phenomenon presents itself, we reach a scientific goal beyond which we cannot pass. When we know that water, with all its properties, results from combining oxygen and hydrogen in certain proportions, we know everything we can know about it; and that corresponds to the *how* and not to the *why* of things. We know how water can be made; but why does the combination of one volume of oxygen with two volumes of hydrogen produce water? We have no idea. In medicine it is equally absurd to concern one's self with the question "why." Yet physicians ask it often. It was probably to make fun of this tendency, which results from lack of the sense of limits to our learning, that Molière put the following answer into the mouth of his candidate for the medical degree. Asked why opium puts people to sleep, he answered: *"Quia est in eo virtus dormitiva, cujus est natura sensus assoupire."* This answer seems ludicrous and absurd; yet no other answer could be made. In the same way, if we wished to answer the question: "Why does hydrogen, in combining with oxygen, produce water?" we should have to answer: "Because hydrogen has the quality of being able to beget water." Only the question "why," then, is really absurd, because it necessarily involves a naïve or ridiculous answer. So we had better recognize that we do not know; and that the limits of our knowledge are precisely here.

In physiology, if we prove, for instance, that carbon monoxide is deadly when uniting more firmly than oxygen with the hemoglobin, we know all that we can know about the cause of death. Experience teaches us that a part of the mechanism of life is lacking; oxygen can no longer enter the organism, because it cannot displace the carbon monoxide in its union with the hemoglobin. But why has carbon monoxide more affinity than oxygen for this substance? Why is entrance of oxygen into the organism necessary to life? Here is the limit of our knowledge in our present state of learning; and even assuming that we succeed in further advancing our experimental analysis, we shall reach a blind cause at which we shall be forced to stop, without finding the primal reason for things.

Let us add that, when the relative determinism of a phenomenon is established, our scientific goal is reached. Experimental analysis of the conditions of the phenomenon, when pushed still further, gives us fresh information, but really teaches us nothing about the nature of the phenomenon originally determined. The conditions necessary to a phenomenon teach us nothing about its nature. When we know that physical and chemical contact between the blood and the cerebral nerve cells is necessary to the production of intellectual phenomena, that points to conditions, but it cannot teach us anything about the primary nature of intelligence. Similarly, when we know that friction and that chemical action produce electricity, we are still ignorant of the primary nature of electricity.

We must therefore, in my opinion, stop differentiating the phenomena of living bodies from those of inorganic bodies, by a distinction based on our own ability to know the nature of the former and our inability to know that of the latter. The truth is that the nature or very essence of phenomena, whether vital or mineral, will always remain unknown. The essence of the simplest mineral phenomenon is as completely unknown to chemists and physicists to-day as is the essence of intellectual phenomena or of any other vital phenomenon to physiologists. That, moreover, is easy to apprehend; knowledge of the inmost nature or the absolute, in the simplest phenomenon, would demand knowledge of the whole universe; for every phenomenon of the universe is evidently a sort of radiation from that universe to whose harmony it contributes. In living bodies absolute truth would be still harder to attain; because, besides implying knowledge of the universe outside a living body, it would also demand complete knowledge of the organism which, as we have long been saying, is a little world (microcosm) in the great universe (macrocosm). [Absolute knowledge could, therefore, leave nothing outside itself; and only on condition of knowing everything could man be granted its attainment. Man behaves as if he were destined to reach this absolute knowledge; and the incessant *why* which he puts to nature proves it. Indeed, this hope, constantly disappointed, constantly reborn, sustains and always will sustain successive generations in the passionate search for truth.

Our feelings lead us at first to believe that absolute truth must lie within our realm; but study takes from us, little by little, these chimerical conceits. Science has just the privilege of teaching us what we do not know, by replacing feeling with reason and experience and clearly showing us the present boundaries of our knowledge. But by a marvellous compensation, science, in humbling our pride, proportionately increases our power. Men of science who carry experimental analysis to the point of relatively determining a phenomenon doubtless see clearly their own ignorance of the phenomenon in its primary cause; but they have become its master; the instrument at work is unknown, but they can use it. This is true of all experimental sciences in which we can reach only relative or partial truths and know phenomena only in their necessary conditions. But this knowledge is enough to broaden our power over nature. Though we do not know the essence of phenomena, we can produce or prevent their appearance, because we can regulate their physico-chemical conditions. We do not know the essence of fire, of electricity, of light, and still we regulate their phenomena to our own advantage. We know absolutely nothing of the essence even of life; but we shall nevertheless regulate vital phenomena as soon as we know enough of their necessary conditions. Only in living bodies these conditions are much more complex and more difficult to grasp than in inorganic bodies; that is the whole difference.

To sum up, if our feeling constantly puts the question *why,* our reason shows us that only the question *how* is within our range; for the moment, then, only the question *how* concerns men of science and experimenters. If we cannot know *why* opium and its alkaloids put us to sleep, we can learn the mechanism of sleep and know *how* opium or its ingredients puts us to sleep; for sleep takes place only because an active substance enters into contact with certain organic substances which it changes. Learning these changes will give us the means of producing or preventing sleep, and we shall be able to act on the phenomenon and regulate it at pleasure.

In the knowledge that we acquire, we should distinguish between two sets of notions: the first corresponds to the *cause* of phenomena, the second to the *means* of producing them. By the cause of a phenomenon we mean the constant and definite condition necessary to existence; we call this the relative determinism or the *how* of things, i.e., the immediate or determining cause. The means of obtaining phenomena are the varied processes by whose aid we may succeed in putting in action the single determining cause which produces the phenomenon. The necessary cause in the formation of water is the combination of two volumes of hydrogen with one of oxygen; this is the single cause which always determines the phenomenon. We cannot conceive of water apart from this essential condition. Subordinate conditions or processes in the formation of water may be extremely varied; only all these processes reach the same result, viz., combination of oxygen and hydrogen in invariable proportions. Let us take another example. I assume that we wish to transform starch into glucose; we have any number of means or processes for doing this, but fundamentally there will always be the identical cause, and a single determinism will beget the phenomenon. This cause is fixation of one more unit of water in the substance, to bring about its transformation. Only we may produce this hydration in any number of conditions and by any number of methods: by means of acidulated water, of heat, of animal or vegetable enzymes; but all these processes finally come to a single condition, hydrolysis of the starch. The determinism, i.e., the cause of the phenomenon, is therefore single, though the means for making it appear may be multiple and apparently very various. It is most important to establish this distinction especially in medicine, where the greatest confusion reigns, precisely because physicians recognize a multitude of causes for the same disease. To convince ourselves of what I am urging we have only to open a treatise on pathology. By no means all the circumstances enumerated are causes; at most they are means or processes by which a disease can be produced. But the real and effective cause of a disease must be *constant* and *determined*, that is unique; anything else would be a denial of science in medicine. It is true that determining causes are much harder to recognize and define in the phenomena of living beings; but they exist nevertheless, in spite of the seeming diversity of means employed. Thus in certain toxic phenomena we see different poisons lead to one cause and to a single determinism for the death of histological units, for example, the coagulation of muscular substance. In the same way, varied circumstances producing the same disease must all correspond to a single and determined pathogenic action. In a word, determinism which insists on identity of effect bound up with identity of cause is an axiom of science which can no more be transgressed in the sciences of life than in the sciences of inorganic matter.

Cause is single thought but means can be thought of variable

X. In the Sciences of Living Bodies, as in Those of Inorganic Bodies, Experimenters Create Nothing; They Simply Obey the Laws of Nature

We know the phenomena of nature only through their relations with the causes which produce them. Now the *law* of phenomena is nothing else than this relation numerically established, in such a way as to let us foresee the ratio of cause to effect in any given case. This ratio, established by observation, enables astronomers to predict celestial phenomena; this same ratio, established by observation and experiment, again enables physicists, chemists, physiologists, not only to predict the phenomena of nature, but even to modify them at pleasure and to a certainty, provided they do not swerve from the ratio which experience has pointed out, i.e., the law. In other terms, we can guide natural phenomena only by submitting to laws that govern them.

Observers can only observe natural phenomena; experimenters can only modify them; it is not given them to create or to destroy them utterly, because they cannot change natural law. We have often repeated that experimenters act, not on phenomena themselves, but on the physico-chemical conditions necessary to their appearance. Phenomena are just the actual expression of the ratio of these conditions; hence, when conditions are similar, the ratio is constant and the phenomenon identical, and when conditions change, there is another ratio, and a different phenomenon. In a word, to make a new phenomenon appear, experimenters merely bring new conditions to pass, but they create nothing, either in the way of force or of matter. At the end of the last century science proclaimed a great truth, to wit, that with respect to matter, nothing is lost, neither is anything created in nature; the bodies whose qualities ceaselessly vary under our eyes are all only transmutations of aggregations of matter of equal weight. In recent times science has proclaimed a second truth which it is still seeking to prove and which in some sense is truly complementary to the first, to wit, that with respect to forces nothing is lost and nothing created in nature; it follows that all the infinitely varied forms of phenomena in the universe are only equivalent transformation of forces, one into another. I reserve for treatment elsewhere the question whether differences separate the forces of living bodies from those of inert bodies; let it suffice for the moment to say that the two preceding truths are universal, and that they embrace the phenomena of living bodies as well as those of inert bodies.

All phenomena, to whatever order they belong, exist implicitly in the changeless laws of nature; and they show themselves only when their necessary conditions are actualized. The bodies and beings on the surface of our earth express the harmonious relation of the cosmic conditions of our planet and our atmosphere with the beings and phenomena whose existence they permit. Other cosmic conditions would necessarily make another world appear in which all the phenomena would occur which found in it their necessary conditions, and from which would disappear all that could not develop in it. But no matter what infinite varieties of phenomena we conceive on the earth, by placing ourselves in thought in all the cosmic conditions that our imagination can bring to birth, we are still forced to admit that this would all take place according to the laws of physics, chemistry and physiology, which have existed without our knowledge from all eternity; and that whatever happens, nothing is created by way either of force or of matter; that only different relations will be produced and through them creation of new beings and phenomena.

When a chemist makes a new body appear in nature, he cannot flatter himself with having created the laws which brought it to birth; he produced only the conditions which the creative law demanded for its manifestation. The case of organic bodies is the same. Chemists and physiologists, in their experiments, can make new beings appear only by obeying the laws of nature which they cannot alter in any way.

It is not given to man to alter the cosmic phenomena of the whole universe nor even those of the earth; but the advances of science

got to the inner environment and by altering observe change duel environment of exploration enable him to alter the phenomena within his reach. Thus man has already gained a power over mineral nature which is brilliantly revealed in the applications of modern science, still at its dawn. The result of experimental science applied to living bodies must also be to alter vital phenomena, by acting solely on the conditions of these phenomena. But here our difficulties are greatly increased by the delicacy of the conditions of vital phenomena and the complexity and interrelation of all the parts grouped together to form an organized being. This is why man can probably never act as easily on animal or vegetable, as on mineral, species. His power over living beings will remain more limited, especially where they form higher, i.e., more complicated organisms. Nevertheless, the difficulties obstructing the power of physiologists do not pertain to the nature of vital phenomena, but merely to their complexity. Physiologists will first begin by getting at phenomena of vegetables and of animals in easier relations with the outer cosmic environment. It appears, at first sight, as if man and the higher animals must escape from its power to change, because they seem freed from the direct influence of the outer environment. But we know that vital phenomena in man, as in the animals nearest him, are connected with the physico-chemical conditions of an inner organic environment. This inner environment we must first seek to know, because this must become the real field of action for physiology and experimental medicine.

CHAPTER II

EXPERIMENTAL CONSIDERATIONS PECULIAR TO LIVING BEINGS

I. The Phenomena of Living Beings Must Be Considered as a Harmonious Whole

So far we have been explaining experimental considerations applicable to both living and inorganic bodies; for living bodies the difference consists merely in the greater complexity of phenomena, making experimental analysis and determination of the conditions incomparably harder. But in the behavior of living bodies we must call the reader's attention to their very special interdependence; in the study of vital functions, if we neglected the physiological point of view, even if we experimented skilfully, we should be led to the most false ideas and the most erroneous deductions.

We saw in the last chapter that the object of the experimental method is to reach the determinism of phenomena, no matter of what nature, whether vital or mineral. We know, moreover, that what we call determinism of a phenomenon means nothing else than the determining cause or immediate cause determining the appearance of phenomena. Thus we necessarily obtain the conditions in which the phenomena exist, and on which the experimenter must act to make the phenomena vary. We therefore consider the various expressions above as equivalents; and the word determinism sums them all up.

It is indeed true, as we have said, that life brings absolutely no difference into the scientific experimental method which must be applied to the study of physiological phenomena, and that in this respect physiological science and physico-chemical science rest on exactly the same principles of investigation. But still we must recognize that determinism in the phenomena of life is not only very complex, but that it is at the same time harmoniously graded. Thus complex physiological phenomena are made up of a series of simpler phenomena each determining the other by associating together or combining for a common final object. Now the physiologist's prime object is to determine the elementary conditions of physiological phenomena and to grasp their natural subordination, so as to understand and then to follow the different combinations in the varied mechanism of animal organisms. The ancient emblem representing life as a closed circle, formed by a serpent biting its own tail, gives a fairly accurate picture of things. In complex organisms the organism of life actually forms a closed circle, but a circle which has a head and a tail in this sense, that vital phenomena are not all of equal importance, though each in succession completes the vital circle. Thus the muscular and nervous organs sustain the activity of the organs preparing the blood; but the blood in its turn nourishes the organs which produce it. Here is an organic or social interdependence which sustains a sort of perpetual motion, until some disorder or stoppage of a necessary vital unit upsets the equilibrium, or leads to disturbance or stoppage in the play of the animal machine. The problem for experimenting physicians consists, therefore, in finding the simple determinism of an organic disorder, that is to say, in grasping the initial phenomenon which brings all the others in its train through a complex determinism as necessary in character as the initial determinism. This initial determinism is like Ariadne's thread guiding the experimenter in the dark labyrinth of physiological and pathological phenomena, and enabling him to understand how their varied mechanisms are still bound together by absolute determinisms. By examples cited further on, we shall see how a dislocation of the organism or an apparently highly complex disorder may be traced back to an initial simple determinism which later produces more complex determinisms. A case in point is poisoning by carbon monoxide (cf. Part III). I am devoting my whole course at the Collège de France this year to the study of *curare*, not for the sake of the substance itself, but because this study shows us how the simplest single determinism, such as the lesion of a terminal motor nerve, re-echoing successively from all the other vital units, leads to secondary determinisms which grow more and more complicated till death ensues. I wish thus to establish experimentally the existence of intra-organic determinisms to which I shall later return, because I consider study of them the true basis of pathology and of scientific therapeutics.

Physiologists and physicians must never forget that a living being is an organism with its own individuality. Since physicists and chemists cannot take their stand outside the universe, they study bodies and phenomena in themselves and separately without necessarily having to connect them with nature as a whole. But physiologists, finding themselves, on the contrary, outside the animal organism which they see as a whole, must take account of the harmony of this whole, even while trying to get inside, so as to understand the mechanism of its every part. The result is that physicists and chemists can reject all idea of final causes for the facts that they observe; while physiologists are inclined to acknowledge an harmonious and pre-established unity in an organized body, all of whose partial actions are interdependent and mutually generative. We really must learn, then, that if we break up a living organism by isolating its different parts, it is only for the sake of ease in experimental analysis, and by no means in order to conceive them separately. Indeed when we wish to ascribe to a physiological quality its value and true significance, we must always refer it to this whole, and draw our final conclusion only in relation to its effects in the whole. It is doubtless because he felt this necessary interdependence among all parts of an organism, that Cuvier said that experimentation was not applicable to living beings, since it separated organized parts which should remain united. For the same reason, other physiologists or physicians, called vitalists, have proscribed and still proscribe experimentation in medicine. These views, which have their correct side, are nevertheless false in their general outcome and have greatly hampered the progress of science. It is doubtless correct to say that the constituent parts of an organism are physiologically inseparable one from another, and that they all contribute to a common vital result; but we may not conclude from this that the living machine must not be analyzed as we analyze a crude machine whose parts also have their rôle to play in a whole. With the help of experimental analysis we must transfer physiological functions as much as possible outside the organism; segregation allows us to see and to grasp hidden conditions of the phenomena, so as to follow them later inside the organism and to interpret their vital rôle. Thus we establish artificial digestion and fecundation, so as to know natural digestion and fecundation better. Thanks to their organic self-regulation, we can also detach living tissues, and by means of artificial circulation or otherwise, we can place them in conditions where we can better study their characteristics. We occasionally isolate an organ by using anesthetics to destroy the reactions of its general group; we reach the same result by cutting the nerves leading to a part, but preserving the blood vessels. By means of experimental analysis, I have even transformed warm-blooded animals, as it were, into cold-blooded animals, so as to study better the characteristics of their histological units; I have succeeded in poisoning glands separately and in making them work, by means of dissected nerves, quite apart from the organism. In this last case we can have a gland, at will, in a state, first, of absolute rest, then, of exaggerated action; when both extremes of the phenomenon are known we can later easily grasp all the intervening stages, and we then understand how a completely chemical function can be regulated by a nervous system, so as to supply organic fluids in conditions that are always the same. We will not further amplify these suggestions about experimental analysis; we sum up by saying that proscribing experimental analysis of organs means arresting science and denying the experimental method; but, on the other hand, that practising physiological analysis, while losing sight of the harmonious unity of an organism, means misunderstanding the science of life and individuality, and leaving it characterless.

After carrying out an analysis of phenomena, we must therefore always reconstruct our physiological synthesis, so as to see the joint action of all the parts we have isolated. À propos of the phrase physiological synthesis, we must further explain our thought. It is generally agreed that synthesis reunites what analysis has divided, and that synthesis therefore verifies analysis, of which it is merely the counterproof or necessary complement. This definition is entirely true for analysis and synthesis of matter. In chemistry, synthesis produces, weight for weight, the same body made up of identical elements combined in the same proportions; but in the case of analyzing and synthesizing the properties of bodies, i.e., synthesizing phenomena, it is much harder. Indeed, the properties of bodies result not merely from the nature and proportions of matter, but also from the arrangement of matter. Moreover, as we know, it happens that properties, which appear and disappear in synthesis and analysis, cannot be considered as simple addition or pure subtraction of properties of the constituent bodies. Thus, for example, the properties of oxygen and hydrogen do not account for the properties of water, which result nevertheless from combining them.

I do not intend to go into these difficult yet fundamental problems about the relative properties of combined or combining bodies; they will find their proper place elsewhere. I shall here only repeat that phenomena merely express the relations of bodies, whence it follows that, by dissociating the parts of a whole, we must make phenomena cease if only because we destroy the relations. It follows also, in physiology, that analysis, which teaches us the properties of isolated elementary parts, can never give us more than a most incomplete ideal synthesis; just as knowing a solitary man would not bring us knowledge of all the institutions which result from man's association, and which can reveal themselves only through social life. In a word, when we unite physiological elements, properties appear which were imperceptible in the separate elements. We must therefore always proceed experimentally in vital synthesis, because quite characteristic phenomena may result from more and more complex union or association of organized elements. All this proves that these elements, though distinct and self-dependent, do not therefore play the part of simple associates; their union expresses more than addition of their separate properties. I am persuaded that the obstacles surrounding the experimental study of psychological phenomena are largely due to difficulties of this kind; for despite their marvellous character and the delicacy of their manifestations, I find it impossible not to include cerebral phenomena, like all other phenomena of living bodies, in the laws of scientific determinism.

Physiologists and physicians must therefore always consider organisms as a whole and in detail at one and the same time, without ever losing sight of the peculiar conditions of all the special phenomena whose resultant is the individual. Yet particular facts are never scientific; only generalization can establish science. But here we must avoid a double stumbling block; for if excess of detail is anti-scientific, excessive generalization creates an ideal science no longer connected with reality. This stumbling block, unimportant to a contemplative naturalist, is large for physicians who must first of all seek objective, practical truths. We must doubless admire those great horizons dimly seen by the genius of a Goethe, an Oken, a Carus, a Geoffroy Saint-Hilaire, a Darwin, in which a general conception shows us all living beings as the expression of types ceaselessly transformed in the evolution of organisms and species,—types in which every living being individually disappears like a reflection of the whole to which it belongs.   In medicine we can also rise to the most abstract generalizations, whether we take the naturalist's point of view and conceive diseases as morbid species to be classified nosologically, or whether we start from the physiological point of view and consider that disease does not exist, in the sense that it is only a special case of a general physiological state.   Doubtless all these brilliant views do, after a fashion, guide and serve us.   But if we gave ourselves up exclusively to hypothetical contemplation, we should soon turn our backs on reality; and in my opinion, we should misunderstand true scientific philosophy, by setting up a sort of opposition or exclusion between practice, which requires knowledge of particulars, and generalizations which tend to mingle all in all.

A physician, in fact, is by no means physician to living beings in general, not even physician to the human race, but rather, physician to a human individual, and still more physician to an individual in certain morbid conditions peculiar to himself and forming what is called his idiosyncrasy.   Hence it seems to follow that medicine, in contrast with other sciences, should be established more and more on particulars.   This opinion is incorrect and based only on appearances; for in all sciences, generalization leads to the law of phenomena and the true scientific goal.   Only we must recognize that all the morphological generalizations to which we alluded above are too superficial and are therefore insufficient for physiologists and physicians.   Naturalists, physiologists and physicians have wholly different problems in view; their investigations advance in far from parallel lines; hence we cannot, for instance, exactly superpose a physiological scale on the geological scale.   Physiologists and physicians delve much more deeply than zoölogists into the problem of biology; physiologists consider the general conditions necessary to vital phenomena as well as the various changes to which they may be subject.   But physicians cannot content themselves with knowing that all vital phenomena occur in identical conditions among all human beings; they must go still further by studying the details of these conditions in each individual considered in given morbid conditions.   Only after delving, then, as deeply as possible into the secrets of vital phenomena in the normal and pathological states can physiologists and physicians attain illuminating and fertile generalizations.

The primary essence of life is a developing organic force, the force which constituted the mediating nature of Hippocrates and the *archeus faber* of Van Helmont. But whatever our idea of the nature of this force, it is always exhibited concurrently and parallel with the physico-chemical conditions proper to vital phenomena. Through study, then, of physico-chemical details, physicians will learn to understand individualities as special cases included in a general law, and will discover there, as everywhere, an harmonious generalization of variety in unity. But since physicians deal with variety, they must always seek to define it in their studies and to comprehend it in their generalizations.

If I had to define life in a single phrase, I should clearly express my thought by throwing into relief the one characteristic which, in my opinion, sharply differentiates biological science. I should say: life is creation. In fact, a created organism is a machine which necessarily works by virtue of the physico-chemical properties of its constituent elements. To-day we differentiate three kinds of properties exhibited in the phenomena of living beings: physical properties, chemical properties and vital properties. But the term "vital properties" is itself only provisional; because we call properties vital which we have not yet been able to reduce to physico-chemical terms; but in that we shall doubtless succeed some day. So that what distinguishes a living machine is not the nature of its physico-chemical properties, complex as they may be, but rather the creation of the machine which develops under our eyes in conditions proper to itself and according to a definite idea which expresses the living being's nature and the very essence of life.

When a chicken develops in an egg, the formation of the animal body as a grouping of chemical elements is not what essentially distinguishes the vital force. This grouping takes place only according to laws which govern the chemico-physical properties of matter; but the guiding idea of the vital evolution is essentially of the domain of life and belongs neither to chemistry nor to physics nor to anything else. In every living germ is a creative idea which develops and exhibits itself through organization. As long as a living being persists, it remains under the influence of this same creative vital force, and death comes when it can no longer express itself; here as everywhere, everything is derived from the idea which alone creates and guides; physico-chemical means of expression are common to all natural phenomena and remain mingled, pell-mell, like the letters of the alphabet in a box, till a force goes to fetch them, to express the most varied thoughts and mechanisms. This same vital idea preserves beings, by reconstructing the living parts disorganized by exercise or destroyed by accidents or diseases. To the physico-chemical conditions of this primal development, then, we must always refer our explanation of life, whether in the normal or the pathological state. We shall see, indeed, that physiologists and physicians can really act only indirectly through animal physico-chemistry, that is to say, through physics and chemistry worked out in the special field of life, where the necessary conditions of all phenomena of living organisms develop, create and support each other according to a definite idea and obedient to rigorous determinisms.

II. Experimental Practice with Living Beings

As we have said, the experimental method and the principles of experimentation are identical for the phenomena of inorganic bodies and the phenomena of living bodies. But it cannot be the same with experimental practice, and it is easy to conceive that the peculiar organization of living bodies requires special processes for its analysis and must offer difficulties *sui generis*. However, the considerations and special precepts, which we shall present to physiologists, to forearm them against sources of error in experimental practice, have to do only with the delicacy, mobility and fugitiveness of vital qualities and the complexity of the phenomena of life. Physiologists, indeed, have only to take apart the living machine, and with the help of tools and processes borrowed from physics and chemistry, to study and measure the various vital phenomena whose law they seek to discover. Each of the sciences possesses, if not an individual method, at least particular processes; and the sciences, moreover, serve as instruments one for another. Mathematics serves as an instrument for physics, chemistry and biology in different degrees; physics and chemistry serve as powerful instruments for physiology and medicine. In the mutual service which sciences render one another, we must of course distinguish between the men of science, who use, and those who carry forward each science. Physicists and chemists are not mathematicians because they make calculations; physiologists are not chemists or physicists because they make use of chemical reagents or physical instruments, any more than chemists and physicists are physiologists because they study the composition or properties of certain animal or vegetable fluids or tissues.   Each science has its problem and its point of view which we may not confuse without risk of leading scientific investigation astray.   Yet this confusion has often occurred in biological science which, because of its complexity, needs the help of all the other sciences.   We have seen, and we still often see chemists and physicists who, instead of confining themselves to the demand that living bodies furnish them suitable means and arguments to establish certain principles of their own sciences, try to absorb physiology and reduce it to simple physico-chemical phenomena.   They offer explanations or systems of life which tempt us at times by their false simplicity, but which harm biological science in every case, by bringing in false guidance and inaccuracy which it then takes long to dispel.   In a word, biology has its own problem and its definite point of view; it borrows from other sciences only their help and their methods, not their theories. This help from other sciences is so powerful that, without it, the development of the science of vital phenomena would be impossible. Previous knowledge of the physico-chemical sciences is therefore decidedly not, as is often said, an accessory to biology, but, on the contrary, is essential to it and fundamental.   That is why I think it proper to call the physico-chemical sciences allied sciences, and not sciences accessory to physiology.   We shall see that anatomy is also a science allied to physiology, just as physiology itself, which requires the help of anatomy and of all the physico-chemical sciences, is the science most closely allied to medicine and forms its true scientific foundation.

The application of physico-chemical sciences to physiology and the use of their processes as instruments, suited to the analysis of the phenomena of life, present a great many difficulties inherent, as we have said, in the mobility and fugitiveness of vital phenomena. The spontaneity and mobility enjoyed by living beings make the properties of organized bodies very hard to fix and to study.   We must return for an instant here to the nature of these difficulties, as I have already had occasion to do in my lectures.[1]

[1] Claude Bernard, *Leçons sur les propriétés physiologiques des altérations pathologiques des liquides de l'organisme.*   Paris, 1859, Vol. I.   *Leçon d'ouverture,* Dec. 9, 1857.

A living body differs essentially from an inorganic body from the point of view of the experimenter. An inorganic body has no sort of spontaneity; as its properties are in equilibrium with outside conditions, it soon settles into physico-chemical indifference, i.e., into stable equilibrium with its surroundings. Hence, all the phenomenal changes that it experiences will necessarily come from alterations occurring in surrounding circumstances; and we can easily see that by taking strict account of these circumstances, we can be sure of having the experimental conditions necessary to a good experiment. A living body, especially in the higher animals, never falls into chemico-physical indifference to the outer environment; it has ceaseless motions, an organic evolution apparently spontaneous and constant; and though this evolution requires outer circumstances for its manifestation, it is nevertheless independent in its advance and modality. As proof of this, we see living beings born, develop, fall ill and die, without the conditions of the outer world changing for the observer.

It follows that, in experimenting on inorganic bodies with the help of such instruments as the barometer, thermometer and hygrometer, we can put ourselves in identical conditions and consequently carry on well-defined and similar experiments. Physiologists and physicians have rightly imitated the physicists and have sought to make their experiments more accurate by using the same instruments. But we can see at once that outer conditions whose changes are of such importance to physicists and chemists are of much less value to physicians. Alterations in the phenomena of inorganic bodies are, in fact, always brought about by an outer cosmic change, and it happens at times that a very slight alteration in the surrounding temperature or in barometric pressure leads to important changes in the phenomena of inorganic bodies. But in man and in the higher animals the phenomena of life may alter without any appreciable outer cosmic change, and slight thermometric or barometric changes often exert no real influence on vital manifestations; and though we cannot say that these outer cosmic influences are essentially nil, circumstances occur where it would be almost ludicrous to take account of them. Such was the experimenter's case who repeated my experiments on puncture of the floor of the fourth ventricle, to cause artificial diabetes: he thought that he exhibited greater accuracy in carefully noting the barometric pressure at the moment of making the experiment.

However, instead of experimenting on man and the higher animals, if we experiment on lower living beings, whether animal or vegetable, we shall see that the thermometric, barometric and hygrometric data, which were so unimportant in the first case, must on the contrary be kept very seriously under consideration in the second. Indeed, if we vary the conditions of humidity, heat and atmospheric pressure for infusoria, the vital manifestations in these beings are altered or annihilated according to the more or less significant variations that we make in the cosmic influences cited above. In vegetables and in cold-blooded animals, the conditions of temperature and humidity in the cosmic environment again play a very large part in the manifestations of life. This is what is called the influence of the seasons, which is familiar to everyone. In fine, then, only the warm-blooded animals and man seem to escape cosmic influences and to have free and independent manifestations. We have already said elsewhere that this kind of independence of vital manifestations in man and the higher animals results from greater perfection of their organism, but does not prove that manifestations of life in these physiologically more perfect beings are subject to other laws or other causes. We know, in fact, that the histological units of our organs express the phenomena of life; now if the functions of these units show no variations under the influence of variations in the temperature, humidity and pressure of the outer atmosphere, it is because they are immersed in an organic environment whose degrees of temperature, humidity and pressure do not change with variations in the cosmic environment. Hence, we must conclude that fundamentally manifestations of life in warm-blooded animals and in man are equally subject to exact and definite physico-chemical conditions.

In recapitulating all that we have already said, we see that conditions of environment in all natural phenomena govern their phenomenal manifestations. The conditions of our cosmic environment generally govern the mineral phenomena occurring on the surface of the earth; but organized beings include within themselves the condition peculiar to their vital manifestations, and in proportion as the organism, i.e., the living machine, perfects itself, and its organized units grow more delicate, it creates conditions peculiar to an organic environment which becomes more and more isolated from the cosmic environment. We thus come back to the distinction which I established long since, and which I believe very fruitful, to wit, that two environments must be considered in physiology: the general macrocosmic environment and the microcosmic environment peculiar to living beings; the latter is more or less independent of the former, according to the degree of perfection of the organism. Moreover, we easily understand what we see here in the living machine, since the same thing is true of the inanimate machines created by man. Thus, climatic changes have no influence at all on the action of a steam engine, though everyone knows that exact conditions of temperature, pressure and humidity inside the machine govern all its movements. For inanimate machines we could therefore also distinguish between a macrocosmic environment and a microcosmic environment. In any case, the perfection of the machine consists in being more and more free and independent, so as to be less and less subject to the influence of the outer environment. The human machine is the more perfect, the better it defends itself from penetration by the influences of the outer environment; as the organism grows old and enfeebled, it becomes more sensitive to the outer influences of cold, heat, humidity, and in general to all other climatic influences.

To sum up, if we wish to find the exact conditions of vital manifestations in man and the higher animals, we must really look, not at the outer cosmic environment, but rather at the inner organic environment. Indeed, as we have often said, it is in the study of these inner organic conditions that direct and true explanations are to be found for the phenomena of the life, health, sickness and death of the organism. From the outside, we see only the resultant of all the inner activities of the body, which therefore seem like the result of a distinct vital force in only the most distant relations with the physico-chemical conditions of the outer environment, and manifesting itself always as a sort of organic personality endowed with specific tendencies. We have elsewhere said that ancient medicine considered the influence of the cosmic environment, of water, air and locality; we may indeed find useful suggestions here as to hygienic and as to morbid changes. But modern experimental medicine will be distinguished for being especially founded on knowledge of the inner environment where normal and morbid as well as medicinal influences take action. But how are we to know this inner

Men have felt this truth in all ages; and in medicine, from the earliest times, men have performed not only therapeutic experiments but even vivisection.   We are told that the kings of Persia delivered men condemned to death to their physicians, so that they might perform on them vivisections useful to science.   According to Galen, Attalus III (Philometor), who reigned at Pergamum, one hundred thirty-seven years before Jesus Christ, experimented with poisons and antidotes on criminals condemned to death.[2]   Celsus recalls and approves the vivisection which Herophilus and Erasistratus performed on criminals with the Ptolemies' consent.   It is not cruel, he says, to inflict on a few criminals, sufferings which may benefit multitudes of innocent people throughout all centuries.[3]   The Grand Duke of Tuscany had a criminal given over to the professor of anatomy, Fallopius, at Pisa, with permission to kill or dissect him at pleasure.   As the criminal had a quartan fever, Fallopius wished to investigate the effects of opium on the paroxysms.   He administered two drams of opium during an intermission; death occurred after the second experiment.[4]   Similar instances have occasionally recurred, and the story is well known of the archer of Meudon [5] who was pardoned because a nephrotomy was successfully performed on him.   Vivisection of animals also goes very far back.   Galen may be considered its founder.   He performed his experiments especially on monkeys and on young pigs and described the instruments and methods used in experimenting.   Galen performed almost no other kind of experiment than that which we call disturbing experiments, which consist in wounding, destroying or removing a part, so as to judge its function by the disturbance caused by its removal.   He summarized earlier experiments and studied for himself the effects of destroying the spinal cord at different heights, of perforating the chest on one side or both sides at once; the effects of section of the nerves leading to the intercostal muscles and of section of the recurrent nerve.   He tied arteries and performed experiments on the mechanism of deglutition.[6]   Since Galen, at long intervals in the midst of

---

[2] Daniel Leclerc, *Histoire de la médecine*, p. 338.

[3] Celsus, *De Medicina*.

[4] Astruc, *De Morbis Venereis*.   Vol. II, pp. 748 and 749.

[5] Rayer, *Traité des maladies des reins*.   Vol. III, p. 213.   Paris, 1841.

[6] Dezeimeris, *Dictionnaire historique*, Vol. II, p. 444.   Daremberg.   *Exposition des connaissances de Galien sur l'anatomie pathologique et la pathologie du système nerveux*.   Thesis, 1841, pp. 13 and 80.

environment of the organism, so complex in man and in the higher animals, unless we go down and, as it were, penetrate into it, by means of experimentation applied to living bodies? That is to say, to analyze the phenomena of life, we must necessarily penetrate into living organisms with the help of the methods of vivisection.

To sum up, only in the physico-chemical conditions of the inner environment can we find the causation of the external phenomena of life. The life of an organism is simply the resultant of all its inmost workings; it may appear more or less lively, or more or less enfeebled and languishing, without possible explanation by anything in the outer environment, because it is governed by the conditions of the inner environment. We must therefore seek the true foundation of animal physics and chemistry in the physico-chemical properties of the inner environment. However, as we shall see further on, it is necessary to consider not only the physico-chemical conditions indispensable to life, but also the peculiar, evolutionary, physiological conditions which are the *quid proprium* of biological science. I have always greatly emphasized this distinction because I believe that it is basic, and that physiological considerations must predominate in a treatise on experimentation applied to medicine. Here indeed we shall find the differences due to influences of age, sex, species, race, or to state of fasting or digestion, etc. That will lead us to consider, in the organism, reciprocal and simultaneous reactions of the inner environment on the organs, and of the organs on the inner environment.

III. Vivisection

We have succeeded in discovering the laws of inorganic matter only by penetrating into inanimate bodies and machines; similarly we shall succeed in learning the laws and properties of living matter only by displacing living organs in order to get into their inner environment. After dissecting cadavers, then, we must necessarily dissect living beings, to uncover the inner or hidden parts of the organisms and see them work; to this sort of operation we give the name of vivisection, and without this mode of investigation, neither physiology nor scientific medicine is possible; to learn how man and animals live, we cannot avoid seeing great numbers of them die, because the mechanisms of life can be unveiled and proved only by knowledge of the mechanisms of death.

medical systems, eminent vivisectors have always appeared. As such, the names of Graaf, Harvey, Aselli, Pecquet, Haller, etc., have been handed down to us. In our time, and especially under the influence of Magendie, vivisection has entered physiology and medicine once for all, as an habitual or indispensable method of study.

The prejudices clinging to respect for corpses long halted the progress of anatomy. In the same way, vivisection in all ages has met with prejudices and detractors. We cannot aspire to destroy all the prejudice in the world; neither shall we allow ourselves here to answer the arguments of detractors of vivisection; since they thereby deny experimental medicine, i.e., scientific medicine. However, we shall consider a few general questions, and then we shall set up the scientific goal which vivisection has in view.

First, have we a right to perform experiments and vivisections on man? Physicians make therapeutic experiments daily on their patients, and surgeons perform vivisections daily on their subjects. Experiments, then, may be performed on man, but within what limits? It is our duty and our right to perform an experiment on man whenever it can save his life, cure him or gain him some personal benefit. The principle of medical and surgical morality, therefore, consists in never performing on man an experiment which might be harmful to him to any extent, even though the result might be highly advantageous to science, i.e., to the health of others. But performing experiments and operations exclusively from the point of view of the patient's own advantage does not prevent their turning out profitably to science. It cannot indeed be otherwise; an old physician who has often administered drugs and treated many patients is more experienced, that is, he will experiment better on new patients, because he has learned from experiments made on others. A surgeon who has performed operations on different kinds of patients learns and perfects himself experimentally. Instruction comes only through experience; and that fits perfectly into the definitions given at the beginning of this introduction.

May we make experiments on men condemned to death or vivisect them? Instances have been cited, analogous to the one recalled above, in which men have permitted themselves to perform dangerous operations on condemned criminals, granting them pardon in exchange. Modern ideas of morals condemn such actions; I completely agree with these ideas; I consider it wholly permissible, however, and useful to science, to make investigations on the properties of tissues immediately after the decapitations of criminals. A helminthologist had a condemned woman without her knowledge swallow larvæ of intestinal worms, so as to see whether the worms developed in the intestines [7] after her death. Others have made analogous experiments on patients with phthisis doomed to an early death; some men have made experiments on themselves. As experiments of this kind are of great interest to science and can be conclusive only on man, they seem to be wholly permissible when they involve no suffering or harm to the subject of the experiment. For we must not deceive ourselves, morals do not forbid making experiments on one's neighbor or on one's self; in everyday life men do nothing but experiment on one another. Christian morals forbid only one thing, doing ill to one's neighbor. So, among the experiments that may be tried on man, those that can only harm are forbidden, those that are innocent are permissible, and those that may do good are obligatory.

Another question presents itself. Have we the right to make experiments on animals and vivisect them? As for me, I think we have this right, wholly and absolutely. It would be strange indeed if we recognized man's right to make use of animals in every walk of life, for domestic service, for food, and then forbade him to make use of them for his own instruction in one of the sciences most useful to humanity. No hesitation is possible; the science of life can be established only through experiment, and we can save living beings from death only after sacrificing others. Experiments must be made either on man or on animals. Now I think that physicians already make too many dangerous experiments on man, before carefully studying them on animals. I do not admit that it is moral to try more or less dangerous or active remedies on patients in hospitals, without first experimenting with them on dogs; for I shall prove, further on, that results obtained on animals may all be conclusive for man when we know how to experiment properly. If it is immoral, then, to make an experiment on man when it is dangerous to him, even though the result may be useful to others, it is essentially moral to make experiments on an animal, even though painful and dangerous to him, if they may be useful to man.

After all this, should we let ourselves be moved by the sensitive

[7] Davaine, *Traité des entozoaires.* Paris, 1860. Synopsis, xxvii.

cries of people of fashion or by the objections of men unfamiliar with scientific ideas? All feelings deserve respect, and I shall be very careful never to offend anyone's. I easily explain them to myself, and that is why they cannot stop me. I understand perfectly how physicians under the influence of false ideas, and lacking the scientific sense, fail to appreciate the necessity of experiment and vivisection in establishing biological science. I also understand perfectly how people of fashion, moved by ideas wholly different from those that animate physiologists, judge vivisection quite differently. It cannot be otherwise. Somewhere in this introduction we said that, in science, ideas are what give facts their value and meaning. It is the same in morals, it is everywhere the same. Facts materially alike may have opposite scientific meanings, according to the ideas with which they are connected. A cowardly assassin, a hero and a warrior each plunges a dagger into the breast of his fellow. What differentiates them, unless it be the ideas which guide their hands? A surgeon, a physiologist and Nero give themselves up alike to mutilation of living beings. What differentiates them also, if not ideas? I therefore shall not follow the example of LeGallois,[8] in trying to justify physiologists in the eyes of strangers to science who reproach them with cruelty; the difference in ideas explains everything. A physiologist is not a man of fashion, he is a man of science, absorbed by the scientific idea which he pursues: he no longer hears the cry of animals, he no longer sees the blood that flows, he sees only his idea and perceives only organisms concealing problems which he intends to solve. Similarly, no surgeon is stopped by the most moving cries and sobs, because he sees only his idea and the purpose of his operation. Similarly again, no anatomist feels himself in a horrible slaughter house; under the influence of a scientific idea, he delightedly follows a nervous filament through stinking livid flesh, which to any other man would be an object of disgust and horror. After what has gone before we shall deem all discussion of vivisection futile or absurd. It is impossible for men, judging facts by such different ideas, ever to agree; and as it is impossible to satisfy everybody, a man of science should attend only to the opinion of men of science who understand him, and should derive rules of conduct only from his own conscience.

The scientific principle of vivisection is easy, moreover, to grasp.

[8] LeGallois, *Œuvres*. Paris, 1824. Preface, p. xxx.

It is always a question of separating or altering certain parts of the living machine, so as to study them and thus to decide how they function and for what. Vivisection, considered as an analytic method of investigation of the living, includes many successive steps, for we may need to act either on organic apparatus, or on organs, or on tissue, or on the histological units themselves. In extemporized and other vivisections, we produce mutilations whose results we study by preserving the animals. At other times, vivisection is only an autopsy on the living, or a study of properties of tissues immediately after death. The various processes of analytic study of the mechanisms of life in living animals are indispensable, as we shall see, to physiology, to pathology and to therapeutics. However, it would not do to believe that vivisection in itself can constitute the whole experimental method as applied to the study of vital phenomena. Vivisection is only anatomical dissection of the living; it is necessarily combined with all the other physico-chemical means of investigation which must be carried into the organism. Reduced to itself, vivisection would have only a limited range and in certain cases must even mislead us as to the actual rôle of organs. By these reservations I do not deny the usefulness or even the necessity of vivisection in the study of vital phenomena. I merely declare it insufficient. Our instruments for vivisection are indeed so coarse and our senses so imperfect that we can reach only the coarse and complex parts of an organism. Vivisection under the microscope would make much finer analysis possible, but it presents much greater difficulties and is applicable only to very small animals.

But when we reach the limits of vivisection we have other means of going deeper and dealing with the elementary parts of organisms where the elementary properties of vital phenomena have their seat. We may introduce poisons into the circulation, which carry their specific action to one or another histological unit. Localized poisonings, as Fontana and J. Müller have already used them, are valuable means of physiological analysis. Poisons are veritable reagents of life, extremely delicate instruments which dissect vital units. I believe myself the first to consider the study of poisons from this point of view, for I am of the opinion that studious attention to agents which alter histological units should form the common foundation of general physiology, pathology and therapeutics. We must always, indeed, go back to the organs to find the simplest explanations of life.

To sum up, dissection is a displacing of a living organism by means of instruments and methods capable of isolating its different parts. It is easy to understand that such dissection of the living presupposes dissection of the dead.

IV. Normal Anatomy in Its Relations with Vivisection

Anatomy is the basis necessary to all medical investigation, whether theoretical or practical. A corpse is an organism deprived of living motion, and the earliest explanation of vital phenomena was naturally sought in dead organs, just as we seek explanation of the action of a machine in motion by studying the parts of the machine at rest. It seems, therefore, that the anatomy of man ought to be the basis of physiology and human medicine. Prejudice, however, opposed dissection of corpses, and in default of the human body, men dissected corpses of animals, in organization as close as possible to man. Thus Galen's anatomy and physiology were done mainly on monkeys. At the same time, Galen also performed dissections of cadavers and experiments on living animals, thus proving that he understood perfectly that dissection of cadavers is significant only in so far as it is compared with dissection of living bodies. In this way, anatomy is indeed only the first step in physiology. Anatomy in itself is a sterile science; its existence is justified only by the presence of living men and animals, well and sick, and by its own possible usefulness to physiology and pathology.

We shall limit ourselves here to considering the kinds of service which anatomy, whether of man or of animals, in our present state of knowledge, can render physiology and medicine. This seems to me the more necessary, because different ideas on this subject hold sway in science; in judging these questions it is, of course, understood that I take the point of view of experimental physiology and medicine, which together make up the truly active science of medicine. In biology we may accept various points of view which establish, as it were, just so many distinct sub-sciences. One science, in fact, is separate from another science only because it has a special point of view and a particular problem. In normal biology, we may distinguish the zoölogical point of view, the direct and comparative anatomical points of view, the special and the general physiological points of view. Zoölogy, describing and classifying species, is only a science of observation used as a vestibule to the true science of animals. The zoölogist merely catalogues animals by outward or inner characteristics of form, according to the types and the laws which nature offers him in the formation of these types. The zoölogist's object is classification of beings according to a sort of plan of creation, and for him the problem is summed up in finding the precise place that an animal should fill in a given classification.

Anatomy, or the science of animal organization, is more closely and necessarily related to physiology. The anatomical point of view differs, however, from the physiological in this, that anatomists wish to explain anatomy by physiology, while physiologists seek to explain physiology by anatomy, which is quite another matter. The anatomical point of view has dominated science from the beginning up to the present, and it still has many partisans. The great anatomists who took this point of view all contributed valiantly, nevertheless, to the development of physiological science; and Haller summed up the idea of the subordination of physiology to anatomy in defining physiology as *anatomia animata*. I can easily understand that the anatomical principle was destined necessarily to present itself first, but I believe that it is false in its limitations, and that it has to-day grown harmful to physiology, after having rendered great service, which I should be the last to deny. Anatomy, in fact, is a simpler science than physiology and consequently should be subordinate to it, instead of dominating it. Every explanation of vital phenomena, based exclusively on anatomical considerations, is necessarily incomplete. The great Haller, who summed up the anatomical period of physiology in his immense and admirable writings, restricted himself so far that his physiology is reduced to an irritable fibre and a sensitive fibre. The whole humoral or physico-chemical side of physiology, which cannot be approached by dissection and which treats of what we call our inner environment, was neglected. The reproach which I am making here against the anatomists who wished to subordinate physiology to their point of view, I make in the same way against chemists and physicists who wish to do the same thing. They are also wrong in endeavoring to subordinate physiology, a more complex science, to chemistry or physics, which are simpler sciences. This has not prevented great service being rendered to physiology by much work in physiological chemistry and physics, even though conceived from a false point of view.

In a word, I consider that the most complex of all sciences, physiology, cannot be completely explained by anatomy. To physiology, anatomy is only an auxiliary science, the most immediately necessary, I agree, but insufficient alone, unless we wish to assume that anatomy includes everything, and that the oxygen, chloride of sodium and iron found in the body are anatomical units of the body. Attempts of this kind have been revived in our day by eminent anatomists and histologists. I do not share these views, because they seem to me to create confusion in the sciences and to lead to darkness instead of light.

Anatomists, we said above, try to invert the true method of explanation, i.e., they take anatomy as an exclusive starting point, and propose to deduce directly from it all the functions solely by logic and without experiments. I have already protested against the pretentiousness of anatomical deductions,[9] by showing that they rest on an illusion of which anatomists are unaware. In anatomy, we must in fact distinguish between two classes of things: (1) The passive mechanical arrangements of various organs and apparatus which, from this point of view, are really nothing but instruments of animal mechanics; (2) the activity of vital units which put in play this diverse apparatus. The anatomy of corpses can certainly take account of the mechanical arrangements of the animal organism; inspection of the skeleton certainly shows a combination of levers whose action we understand solely through their arrangement. So with the system of canals or of tubes conducting fluids; and thus the valves in the veins have mechanical functions which put Harvey on the track leading to the discovery of the circulation of the blood. The reservoirs, the bladders, the various pockets in which secreted and excreted fluids reside, offer mechanical arrangements which more or less clearly indicate the functions which they must fulfill, without our necessarily having recourse to experiment on the living to learn it. But we should notice that these mechanical deductions are by no means absolutely restricted to the functions of living beings; we deduce everywhere, in the same way, that pipes are meant to conduct, reservoirs meant to hold and levers meant to move.

When we come to the active or vital elements which put all the passive instruments of the organism in play, then anatomy of corpses

[9] Cf. Claude Bernard, *Leçons de physiologie expérimentale*, Paris, 1856, Vol. II. *Leçon d'ouverture*, May 2, 1855.

cannot and does not teach anything. All our knowledge on this subject must necessarily come from experiment or from observation of the living; when, therefore, anatomists believe that they are making deductions solely from anatomy and without experiments, they forget that their starting point was the same experimental physiology which they seem to disdain. When anatomists deduce the functions of an organism, as they say, from their texture, they merely use knowledge gained on the living, to interpret what they see in the dead; but anatomy really teaches them nothing, it merely supplies them with the quality of a tissue.

So when anatomists meet with muscular fibres in some part of the body, they infer contractile motion; when they meet gland cells, they infer secretion; when they meet with nerve fibres, they infer sensation or movement. But what taught them that muscular fibre contracts, that gland cells secrete, that a nerve is sensory or motor, unless it was observation of the living, or, in other words, vivisection? Only, noting that these contractile, secreting or nerve tissues have definite anatomical forms, they establish a relation between the form of the anatomical unit and its functions, so that when they meet one, they infer the other. But, I repeat, dead anatomy teaches nothing; it merely leans on what experimental physiology teaches; and a clear proof of this is that, where experimental physiology has learned nothing as yet, anatomists can interpret nothing by anatomy alone. Thus, the anatomy of the spleen, the suprarenal glands and the thyroid is as well known as the anatomy of a muscle or of a nerve, and nevertheless anatomists are silent as to the uses of these parts. But as soon as physiologists have discovered something about the functions of these organs, anatomists will put the physiological properties noted into relation with their anatomical observations. I must also point out that anatomists, in their localizations, can never go beyond the teachings of physiology, except under penalty of falling into error. Thus, if anatomists, on the basis of physiological teaching, suggest that, where muscular fibres are present, there are contraction and movement also, they may not infer that, where they see no muscular fibre, there is never contraction or movement. Experimental physiology has proved, in fact, that contracting units are of various forms, among them some which anatomists have not yet been able to define.

In a word, to know something about the functions of life, you must study them in the living. Anatomy yields only characteristics

by which to recognize tissues, but itself teaches nothing about their
vital properties. How indeed could the form of the nerve cell show
us the nervous properties which it transmits? How could the form
of a liver cell show us that sugar is made in it? How could the form
of a muscle fibre teach us about muscular contraction? We have
here only an empirical relation established by comparative observation of the living and the dead. I remember having often heard
de Blainville try to differentiate in his lectures between what should
be called, according to him, a *substratum,* and what should be called,
on the other hand, an *organ.* In an organ, according to de Blainville,
we should be able to understand the necessary mechanical relation
between a structure and its function. Thus, from the form of bony
levers, he said, we conceive a definite motion; from the disposal of
the blood of the reservoirs for liquids, and of the excretory ducts of
glands, we understand that liquids are put in circulation or retained
by mechanical arrangements that we can explain. But as for the
encephalon, he added, no material relations can be established between
the structure of the brain and the nature of intellectual phenomena.
Therefore, concluded de Blainville, the brain is not the organ of
thought, it is merely a *substratum.* We may accept, if we like, de
Blainville's distinction, but if so, it will be general and not limited
to the brain. Indeed, if we understand that a muscle inserted between two bones may act mechanically as a power drawing them together, we by no means understand how the muscle contracts, and we
can just as well say that the muscle is the substratum of contraction.
Though we understand that a fluid secreted by a gland flows out of its
tubes, we cannot thereby conceive any idea of the essence of secretory
phenomena. And we may just as well say that the gland is a substratum of secretion. To sum up, the anatomical point of view is
wholly subordinate to the point of view of experimental physiology,
as an explanation of vital phenomena. But, as we said above, there
are two things in anatomy: the tools of the organism and the essential agents of life. The essential agents of life depend upon the vital
properties of our tissues, which can be defined only by observation
or by experiment on the living. These agents are the same in all
animals, without distinction of class, genus or species. Here is the
domain of general anatomy and physiology. Next come tools of life,
which are nothing but mechanical tools or weapons with which nature
especially provides each organism according to its class, its genus

or its species. We may even say that the special tools constitute the species; for a rabbit differs from a dog only because one has organs that make it eat grass, and the other organs that force it to eat flesh. But as to the inmost phenomena of life, the two animals are identical. A rabbit is carnivorous if we give him meat ready prepared, and I long ago proved that all fasting animals are carnivorous.

Comparative anatomy is merely an inner zoölogy; its aim is to classify the apparatus or tools of life. These classifications should corroborate or rectify the characteristics suggested by outer forms. Thus the whale, which might be put with the fishes by reason of its outer form, is placed with the mammals because of its interior organization. Comparative anatomy shows us also that the tools of life are arranged in necessary and harmonious relations with the whole organism. Thus, an animal with claws should have the jaws, teeth and the articulations of the limbs disposed in a definite way. The genius of Cuvier amplified these views and derived from them a new science, paleontology, which reconstructs an entire animal from a fragment of his skeleton. The object of comparative anatomy, then, is to show the functional harmony of the tools with which nature has endowed an animal and to teach us the changes necessary in these tools according to various circumstances of animal life. But beneath all these changes comparative anatomy always shows us the uniform plan of creation; thus any number of organs exist, not as aids to life (they are often actually harmful), but as characteristics of the species or as vestiges of a single plan of organic composition. The stag's antlers have no use favorable to the animal's life; the shoulder blade of a slow-worm and the mammæ in males are vestiges of organs that have lost their functions. Nature, as Goethe said, is a great artist; to ornament forms, she often adds organs that are useless to life in itself, as an architect makes ornaments for his building, such as friezes, cornices and volutes which are useless for habitation.

The object of comparative anatomy and physiology is, therefore, to find the morphological laws of the tools and the organs which together make up organisms. Comparative physiology, in so far as it infers functions by comparing organs, would be an insufficient and false science if it rejected experimentation. Comparing the forms of limbs or of the mechanical apparatus of life may suggest the uses of these parts. But what can the form of the liver or the pancreas tell us about the function of these organs? Has not experiment shown the mistake of likening the pancreas to a salivary gland?[10] What can the form of the brain or the nerves teach us about their functions? All that we know has been learned by the observation of the living, or by experiment. What can we say about fishes' brains, for instance, until experiment has clarified the question? In a word, anatomical deduction has yielded what it can. To linger in this path means lagging behind the progress of science and believing that we can impose scientific principles without experimental verification. That, in a word, is a relic of the scholasticism of the Middle Ages. But, on the other hand, comparative physiology, in so far as it relies on experiment and seeks the properties of tissues and organs in animals, does not seem to me to have separate existence as a science. It falls back necessarily into special or general physiology, since its object is the same.

We distinguish between the various biological sciences only by the goal which we set ourselves or the idea which we pursue in studying them. Zoölogists and comparative anatomists see all living beings as a whole, and by studying the outer and inner characteristics of beings, they seek to discover the morphological laws of their evolution and their transformation. Physiologists take a quite different point of view: they deal with just one thing, the properties of living matter and the mechanism of life, in whatever form it shows itself. For them, genus, species and class no longer exist. There are only living beings; and if they choose one of them for study, that is usually for convenience in experimentation. Physiologists also follow a different idea from the anatomists. The latter, as we have seen, try to infer the source of life exclusively from anatomy; they therefore adopt an anatomical plan. Physiologists adopt another plan and follow a different conception; instead of proceeding from the organ to the function, they start from the physiological phenomenon and seek its explanation in the organism. To solve the problem of life, physiologists therefore call to their aid all the sciences,—anatomy physics, chemistry, which are all allies serving as indispensable tools for investigation. We must, therefore, necessarily be familiar enough with these various sciences to know all the resources which may be drawn from them. Let us add, in ending, that from every

[10] Claude Bernard, *Mémoire sur le pancréas.* (*Supplément aux Comptes rendus de l'Académie des sciences.* 1856. Vol. I.)

biological point of view, experimental physiology is in itself the one
active science of life, because by defining the necessary conditions of
vital phenomena it will succeed in mastering them and in governing
them through knowledge of their peculiar laws.

V. Pathological Anatomy and Dissection in Relation to Vivisection

What we said in the last paragraph about normal anatomy and
physiology may be repeated for pathological anatomy and physiology.
We find similarly three points of view appearing one after another,
the taxonomical or nosological point of view, the anatomical point of
view and the physiological point of view. We cannot here go into detailed study of these questions, which would include neither more
nor less than the entire history of medical science. We shall limit
ourselves to suggesting our idea in a few words.

While observing and describing diseases, men must have sought
at the same time to classify them, as they sought to classify animals,
and according to precisely the same principles, by artificial or natural methods. Pinel applied to pathology the natural classification
introduced into botany by de Jussieu, and into zoölogy by Cuvier.
It is sufficient to quote the first sentence of Pinel's *Nosography:*
"Given a disease, to find its place in a nosological scheme." [11]   No
one, I think, will consider this the goal of all medicine; it is merely
a partial point of view, the taxonomic point of view.

After nosology came the anatomical point of view; that is, after
considering diseases as morbid species, men try to place them ana-
tomically. It was thought that, just as there is a normal organization to take account of vital phenomena in the normal state, so there
must be an abnormal organization to take account of morbid phenomena. Though the point of view of pathological anatomy can already be recognized in Morgagni and Bonnet, still it is especially
in this century, under the influence of Broussais and Laennec,
that pathological anatomy has been systematically built up. Men
compared the anatomy of diseases, they classified changes in tissues,
but they also tried to bring these changes into relation with the
morbid phenomena and, as it were, to deduce the second from the first.
The same problems presented themselves as in comparative, normal

---

[11] Pinel, *Nosographie philosophique.*   1800.

anatomy. In the case of morbid changes producing physical or mechanical alteration in a function, as for instance a vascular compres-
sion or mechanical lesion of a limb, men could understand the relation connecting the morbid symptom with its cause and could make
what is called a rational diagnosis. Laennec, one of my predecessors
in the chair of medicine at the Collège de France, immortalized himself in this field by the precision which he gave to physical diagnosis
of diseases of the heart and lungs. But diagnosis became impossible
in the case of diseases where changes were imperceptible with our
present means of investigation. No longer able to find an anatomical
relation, men said then that the disease was essential, i.e., without
any lesion; which is absurd, for it amounts to acknowledging an
effect without a cause. Men came to understand that, to find the
explanation of such diseases, they must carry their investigations
into the minutest parts of the organism where life has its seat. The
new era of microscopic pathological anatomy was inaugurated in
Germany by Johannes Müller; [12] and an illustrious professor in Berlin, Virchow, recently systematized microscopic pathology.[13] So in
changes of the tissues, they found proper characteristics for defining
diseases. A propos of this, they invented the name pathological physiology, to designate pathological function in relation to abnormal
anatomy. I shall not have to consider whether these expressions,
pathological anatomy and physiological pathology, are well chosen.
I shall simply say that the pathological anatomy, whose pathological
phenomena they define, is subject to the same objection of insufficiency that I have already made to normal anatomy. First, the
pathological anatomists assume it proved that anatomical changes
are always primary, which I do not admit, believing the contrary,
that a pathological change is very often secondary and is the consequence or fruit of the disease instead of its germ; which does not
prevent this product from later becoming a morbid germ of other
symptoms. I shall therefore not admit that cells or fibres of tissues
are always primarily attacked; a morbid physico-chemical change in
the organic environment being able, in itself, to lead to the morbid
phenomena, in the manner of a toxic symptom which occurs, without

---

[12] Müller, *De Glandularum secernentium structura penitiori earumque prima
formatione in homine atque animalibus.* Leipzig, 1830.
[13] Virchow, *La Pathologie cellulaire basée sur l'étude physiologique et pathologique des tissus,* translated by P. Picard. Paris, 1860.

primary lesion of the tissues, through mere change in the environment.

The anatomical point of view is therefore insufficient, and the changes noted in cadavers after death really show characteristics by which to recognize and classify diseases, rather than lesions capable of explaining death. It is indeed strange to see how little attention most physicians pay to this latter point of view, which is the true point of view of physiology. When a physician, for example, makes a typhoid autopsy, he notes the intestinal lesions and is satisfied. But in reality that explains absolutely nothing about either the cause of the disease, or the action of drugs, or the reason for death. Microscopic anatomy teaches us no more about it, for when a person dies of tuberculosis or pneumonia or typhoid fever, the microscopic lesions found after death existed before, and often long before; death is not explained by the particles either of the tubercle or of Peyer's patches in the intestines or of other morbid products; death, in fact, can be understood only because some histological unit has lost its physiological properties, a loss which has brought on the disruption of vital phenomena. But to grasp the physiological lesions in their relations with the mechanism of death, we should have to make autopsies on cadavers immediately after death, which is impossible. This, then, is why we must perform experiments on animals and must necessarily give medicine the experimental point of view, if we mean to found a truly scientific medicine which shall logically embrace physiology, pathology and therapeutics. For many years I have done my best to advance in this direction.[14] But the point of view of experimental medicine is most complex, in that it is physiological and also includes explanation of pathological phenomena by anatomy. À propos of pathological anatomy, I shall repeat what I said about normal anatomy, to wit, that anatomy in itself teaches nothing without observation of the living. For pathology we must therefore establish pathological vivisection, that is to say, we must create diseases in animals and sacrifice them at various stages of these diseases. We may also study in the living the changes in the physiological properties of tissues as well as the changes in the cells or the environments. When the animal dies, we must make an autopsy immediately after death, just as if we were dealing with one of those instantaneous diseases called poisoning, for fundamentally

[14] Claude Bernard, *Cours de pathologie expérimentale.* (*Medical Times*, 1860.)

there is no difference in the study of physiological activities, whether morbid, toxic or medicinal. In a word, a physician should not hold to anatomical pathology alone, to explain the disease; he starts from observation of the patient and later explains the disease by physiology with the help of pathological anatomy and all the allied sciences used by investigators of biological phenomena.

VI. THE VARIETY OF ANIMALS SUBJECTED TO EXPERIMENTATION; THE VARIABILITY OF ORGANIC CONDITIONS WHICH THEY PRESENT TO EXPERIMENTERS

All animals may be used for physiological investigations, because, with the same properties and lesions in life and disease, the same result everywhere recurs, though in mechanism the vital manifestations vary greatly. However, the animals most used by physiologists are those procured most easily, and here we must set in the front rank domestic animals such as dogs, cats, horses, rabbits, oxen, sheep, pigs, barnyard fowl, etc., but if we had to reckon up the services rendered to science, frogs would deserve the first place. No other animal has been used for greater or more numerous discoveries, at all points in science; and even to-day, physiology without frogs would be impossible. If the frog, as has been said, is the Job of physiology, that is to say, the animal most maltreated by experimenters, it is certainly the animal most closely associated with their labors and their scientific glory.[15] To the list of animals cited above, we must add many others, warm-blooded and cold-blooded, vertebrates and invertebrates, and even infusoria which may be used for special investigations. But specific diversity is not the sole difference between the animals which physiologists subject to experimentation; in the condition in which they are found, they present many differences which at this point require consideration; for without knowledge or appreciation of their individual characteristics, we can have neither biological exactness nor precision in experimentation.

The first condition for making an experiment is that its circumstances must be so well known and so precisely defined that we can always reconstruct them and reproduce the same phenomena at will. We have said elsewhere that this fundamental condition of

[15] C. Duméril, *Notice historique sur les découvertes faites dans les sciences d'observation par l'étude de l'organisme des grenouilles.* 1840.

experimentation is relatively easy to fulfill in inorganic beings and is surrounded with great difficulties in living beings, particularly in warm-blooded animals. In fact, we must not only reckon with variations in the surrounding cosmic environment, but must also reckon with variations in the organic environment,—that is to say, the present state of the animal organism. We should therefore be greatly in the wrong if we believed it enough, in making an experiment on two animals of the same species, to place them in exactly the same experimental conditions. In every animal, certain physiological conditions of the inner environment have an extreme variability, which, at a given moment, produces appreciable differences, from the experimental point of view, between animals of the same species whose outward appearance is identical. I believe that, more than anyone else, I have emphasized how necessary it is to study physiological conditions, and have shown that knowledge of them is the necessary foundation of experimental physiology.

We must indeed admit that vital phenomena in an animal vary only with precise and definite conditions of the inner environment. We shall therefore try to find these experimental, physiological conditions, instead of tabulating the variations in phenomena and taking averages as expressions of reality; we should thus reach conclusions based on correct statistics, but with no more scientific reality than if they were wholly arbitrary. If we wish to wipe out the diversity evident in organic fluids by taking the averages of all the analyses of urine or blood, even from an animal of the same species, we should thus have a mere combination of these humors corresponding to no definite physiological state of the animal. I have indeed shown that, in fasting, urine always has the same definite composition; I have shown that the blood coming out of an organ is quite different, according to whether the organ is in a state of activity or rest. If we look for sugar in the liver, for instance, and make tabulations of its absence or presence, and take averages to find out how many times per hundred there is sugar or glycogen in that organ, we shall find a number which, whatever it is, means nothing, because, as I have shown, in certain physiological conditions there is always sugar, and in other conditions there is never any. Now taking another point of view, if we meant to consider all the experiments successful in which there was hepatic sugar and consider all those unsuccessful in which there was none, we should fall into another but no less reprehensible kind of error. In fact, I have posited this principle: *there never are any unsuccessful experiments;* they are all successful in their own definite conditions, so that negative cannot nullify positive results. I shall return elsewhere, however, to this important subject. For the moment, I wish merely to call to the attention of experimenters the importance of defining organic conditions, because, as I have already said, they are the one foundation of experimental physiology and medicine. In what follows, a few indications will suffice, since those conditions will be studied later, *à propos* of each particular experiment, from the three points of view: physiological, pathological and therapeutic.

In every experiment on living animals, three kinds of physiological conditions peculiar to the animal must be considered, apart from general cosmic conditions, to wit: anatomical operative conditions, physico-chemical conditions of the inner environment and organic conditions of units in the tissues.

1. *Operative Anatomical Conditions.*—Anatomy is the necessary foundation of physiology, and never can we become good physiologists if we are not first deeply versed in anatomical studies and trained in delicate dissections, so as to be able to make the preparations which are often required for physiological experiments. In fact, operative anatomical physiology has not yet been founded; the zoölogists' comparative anatomy is too vaguely superficial for physiologists to find in it the exact topographical knowledge that they need; the anatomy of domestic animals is done by veterinarians from too special and restricted a point of view to be of great use to experimenters, and thus physiologists are ordinarily reduced to making for themselves the anatomical investigations which they need to devise their experiments. It is evident that, in severing a nerve, tying a duct or injecting a vessel, knowledge of the anatomical arrangement of parts, in the animal operated on, is absolutely indispensable to understanding and defining the physiological results of the experiment. Some experiments would be impossible with certain species of animals, and intelligent choice of an animal offering a happy anatomical arrangement is often a condition essential to the success of an experiment and to the solution of an important physiological problem. Anatomical arrangements may sometimes present anomalies which must also be thoroughly known, as well as the variations observed between one animal and another. In the sequel of this

work, I shall therefore be careful always to describe experimental methods, including the anatomical arrangement, and I shall show that divergencies of opinion among physiologists have been caused, more than once, by anatomical differences which they failed to reckon with, when interpreting the results of experiments. As life is merely a mechanism, anatomical arrangements peculiar to certain animals may seem insignificant at first sight; yet these seemingly futile minutiæ are often enough to change the physiological manifestations completely and to form what we call a highly important idiosyncrasy. A case in point is section of the two facial nerves, which is mortal in horses, but not mortal in other closely related animals.

2. *Physico-chemical Conditions of the Inner Environment.*— Life is made manifest by the action of outer stimuli on irritable living tissues which react by manifesting their special properties. The physiological conditions of life are therefore nothing but the special physico-chemical stimuli which set in action the tissues of the organism. These stimuli are found in the atmosphere or the environment which the animal inhabits; but we know that the properties of the general outer atmosphere pass into the organic atmosphere in which all the physiological conditions of the outer atmosphere are found, plus a certain number of others peculiar to the inner environment. We shall content ourselves here with naming the principal physico-chemical conditions of the inner environment to which experimenters should direct their attention. These, moreover, are only the conditions offered by every environment in which life manifests itself.

Water is the first indispensable condition of every vital manifestation, as of every manifestation of physico-chemical phenomena. In the outer cosmic environment we may distinguish between aquatic and aerial animals; but this distinction can no longer be made between histological units; plunged in the inner environment, they are aquatic in all living beings, that is to say, they are bathed in organic fluids, including very large quantities of water. The proportion of water at times reaches 90 to 99 per cent. in the organic fluids, and when this proportion of water is notably diminished, peculiar physiological troubles result. Thus, if we remove water from frogs by prolonged exposure to dry air, and if we reduce the quantity of water in the blood by introducing into the body substances with a very high endosmotic equivalent, we witness convulsive phenomena, which cease as soon as we restore the normal proportion of water. Complete removal of water from living bodies invariably leads to death in large organisms provided with delicate histological units; but it is well known that in small inferior organisms, removal of water invariably suspends life. Cases in point are the return to life of rotifera, of tardigrades and the small eels in mildewed wheat. There are numberless cases of latent life in vegetables and animals, due to removal of the organism's water.

Temperature has a considerable influence on life. Raising the temperature makes vital phenomena as well as the manifestation of physico-chemical phenomena more active. In the outer cosmic environment, variations of temperature create the seasons which are characterized only by variations in the behavior of animal and vegetable life on the surface of the earth. These variations take place only because the inner environment or organic atmosphere of plants and certain animals remains in equilibrium with the outer atmosphere. If we put plants in hothouses, the influence of winter no longer makes itself felt; the case of cold-blooded and hibernating animals is the same. But warm-blooded animals keep their organic units, as it were, in a hothouse; so they do not feel the influence of hibernation. However, since this is only a special resistance of the inner environment to falling into heat equilibrium with the outer environment, this resistance can be overcome in certain cases, and in some circumstances warm-blooded animals can warm or cool themselves. The maximum temperature compatible with life does not generally rise above 75 degrees. The minimum does not go below the freezing point of the organic animal or vegetable fluids. However, these extremes may vary. In warm-blooded animals the temperature of the inner atmosphere is normally from 38 to 40 degrees: it cannot go above 45 to 50 degrees nor below 15 to 20 degrees, without causing physiological disturbance or even death if the variations are rapid. In hibernating animals, the gradual change of temperature which occurs may go much lower, causing the progressive disappearance of manifestation of life, to the point of lethargy or latent life, which may sometimes last a very long time, if the temperature does not vary.

Air is necessary to the life of all vegetable and animal beings; air therefore exists in the inner organic atmosphere. The three gases of the outer air, oxygen, nitrogen and carbonic acid gas, are in solution in the organic fluids in which the histological units breathe directly, like fish in water. Cessation of life through removal of these gases, especially oxygen, is what we call death by asphyxiation. In living beings, there is a constant interchange of gases between the inner and the outer environment; yet vegetables and animals, as we know, differ in respect to the changes which they cause in the surrounding air.

Pressure exists in the outer atmosphere; we know that on the surface of the earth the air exerts on living beings a pressure that lifts a column of mercury to a height of about 76 centimeters. In the inner atmosphere of warm-blooded animals, the nutritive fluids circulate under the influence of a pressure greater by about 150 mm. than the outer atmospheric pressure, but this does not necessarily indicate that the histological units really support such pressure. The influence of variations of pressure on the manifestations of life of organic units, moreover, is little known. We learn, however, that life cannot appear in too highly rarefied air, not only because the gases of the outer air cannot then dissolve in the alimentary fluid, but also because the gases dissolved in it escape. This is what we observe when we place a small animal under an air pump; its lungs are obstructed by the gases liberated in the blood. The articulates bear this removal of air much better, as various experiments have shown. Fishes in the depths of the sea sometimes live under considerable pressure.

The chemical composition of the cosmic or outer environment is constant and very simple. Its composition is that of the air, which remains the same, except for the proportions of water vapor and a few electric and other conditions which vary. The chemical conditions of the inner or organic environments are much more complex; and complication increases as the animal itself becomes higher and more complex. Organic environments, as we have said, are always aqueous; they hold in solution definite saline and organic substances; they show fixed reactions; the lowest animal has its own organic environment; infusoria have an environment belonging to them, in this sense that they are more permeated than is a fish with the water in which he swims. In the organic environments of the higher animals, the histological units are like veritable infusoria, that is to say, they are provided with an environment proper to themselves which is not the general organic environment. Thus a corpuscle of blood is permeated with a fluid different from the serum in which it floats.

3. *Organic Conditions.*—Organic conditions are those which correspond to the evolution or change of the vital properties of organic units. Variation of these conditions necessarily leads to changes whose principal features we must recall here. Manifestations of life grow more varied, delicate and active in proportion as beings rise in the scale of organization. But susceptibility to disease also increases at the same time. Experimentation, as we have already said, is necessarily more difficult as organization becomes more complex.

Animal and vegetable species are separated by peculiar conditions which prevent them from mingling, so that fecundation, grafting and transfusion cannot be performed between beings of different species. These are problems of the greatest interest, which I believe may be attacked and reduced to differences in the physicochemical properties of environment.

In the same species of animals, breeds may still show a certain number of very interesting differences. In different breeds of dogs and horses I have noted very peculiar physiological characteristics related, in different degrees, to the properties of certain histological units, particularly of the nervous system. Finally, among individuals of the same breed we find physiological peculiarities which are also connected with special variations in the properties of certain histological units. These, therefore, are what we call idiosyncrasies.

The same individual is unlike himself at some periods in his evolution; this leads to differences connected with age. After birth, the phenomena of life are of slight intensity; soon they become very active, to slow down again toward old age.

Sex and the physiological state of the genital organs may lead to changes which are sometimes very profound, especially in the lower animals, among which the physiological characteristics of the larvæ, in certain cases, completely differ from the characteristics of the animal when full grown and endowed with genital organs.

Molting at times leads to such profound organic changes that experiments performed on animals in different stages yield by no means the same results.[16]

Hibernation also leads to great differences in the phenomena of

---

[16] See L. Ziegler, *Ueber die Brunst und den Embryo der Rehe.* Hanover, 1843.

life; and operating on a frog or a toad is by no means the same thing in summer as in winter.[17]

States of digestion or fasting, of health or disease, also cause great changes in the intensity of vital phenomena and so in the resistance of animals to the influence of certain toxic substances and in their susceptibility to one or another parasitic or virulent disease. Habit is yet another condition potent in changing organisms. This condition is one of the most important to keep in mind, especially when we intend to experiment on the action of toxic or medicinal materials.

The size of animals also involves important changes in the intensity of vital phenomena. In general, vital phenomena are more intense in small animals than in large, which means, as we shall see further on, that we cannot rigorously measure physiological phenomena in proportion to weight.

To sum up, after all that we have previously said, we see what huge complexity inheres in animal experimentation because of the numberless factors with which the physiologist must reckon. Nevertheless, we may succeed if we make proper distinctions and subordinations, as we have just said, among the various factors, and if we seek to connect the factors in question with definite physico-chemical conditions.

VII. The Choice of Animals; the Usefulness to Medicine of Experiments on Various Species of Animals

Among the objections that physicians have offered to experimentation is one which must be seriously considered because it throws doubt on the usefulness of animal experiments to human physiology and medicine. It has been said, indeed, that experiments performed on a dog or a frog may be conclusive in their application to dogs and frogs, but never to man, because man has a physiological and pathological nature proper to himself and different from all other animals. It has been further stated that to be really conclusive for man, experiments would have to be made on man or on animals as near to him as possible. It was surely with this idea that Galen chose a monkey for his experiments, and Vesalius a pig, as subjects more closely

[17] See Stannius, *Beobachtungen über Verjüngungsvorgänge im thierischen Organismus.* Rostoch and Schwerin, 1853.

resembling man in his omnivorous capacity. Even to-day, many people choose dogs for experiments, not only because it is easier to procure this animal, but also because they think that experiments performed on dogs can more properly be applied to man than those performed on frogs. How well founded are these opinions? How much importance should we ascribe to the choice of animals in relation to the usefulness of the experiment to physicians?

As far as direct applicability to medical practice is concerned, it is quite certain that experiments made on man are always the most conclusive. No one has ever denied it. Only, as neither the moral law nor that of the state permits making on man the experiments which the interests of science imperatively demand, we frankly acclaim experimentation on animals: from the theoretic point of view, experiments on all sorts of animals are indispensable, while from the immediately practical point of view, they are highly useful to medicine. In fact, as we have already often expressed it, two things must be considered in the phenomena of life: first the fundamental properties of vital units which are general, then arrangements and mechanisms in organizations, which give each animal species its peculiar anatomical and physiological form. Now, among all the animals on which physiologists and physicians may experiment, some are better suited than others to the studies depending on these two points of view. Here we shall merely say in general that, for the study of tissues, cold-blooded animals or young mammals are more appropriate, because the properties of their living tissues vanish more slowly and so can better be studied. There are also experiments in which it is proper to choose certain animals which offer favorable anatomical arrangements or special susceptibility to certain influences. For each kind of investigation we shall be careful to point out the proper choice of animals. This is so important that the solution of a physiological or pathological problem often depends solely on the appropriate choice of the animal for the experiment so as to make the result clear and searching.

General physiology and pathology are necessarily based on the study of tissues in all animals, for a general pathology that did not ultimately rest on considerations drawn from the comparative pathology of animals at all stages of organization could build up a collection of generalities about human pathology, but never a general pathology in the scientific sense of the word. Just as an organism can live only by the normal manifestation of its properties or the help of one or more of its vital units, so the organism can become diseased only by abnormal manifestation of the properties of one or more of its vital units. Now the vital units, being of like nature in all living beings, are subject to the same organic laws. They develop, live, become diseased and die under influences necessarily of like nature, though manifested by infinitely varying mechanisms. A poison or a morbid condition, acting on a definite histological unit, should attack it in like circumstances in all animals furnished with it; otherwise these units would cease to be of like nature; and if we went on considering as of like nature units reacting in different or opposite ways under the influence of normal or pathological vital reagents, we should not only deny science in general, but also bring into zoölogy confusion and darkness that would absolutely block its advance; for the quality which should be placed in the front rank of the science of life and should dominate all the rest is vitality. The vital quality may doubtless offer great diversities in degree and kind of manifestation, according to peculiar circumstances of environment or mechanism shown by healthy or diseased organisms. The lower organisms have fewer distinct vital units than do the higher organisms; whence it follows that these beings are less easily attacked by morbid or mortal influences. But in animals of the same class, order or species there are also constant or variable differences which medical physiologists must absolutely know and explain, because these differences, though resting on delicate distinctions, give phenomena an essentially different aspect. The problem of science will consist precisely in this, to seek the unitary character of physiological and pathological phenomena in the midst of the infinite variety of their particular manifestations. Experimentation on animals is therefore one foundation of comparative physiology and pathology; and we shall later quote examples to prove how important it is not to lose sight of the above ideas.

In special questions of physiology and pathology, experimentation on the higher animals yields daily results, which are applicable in practice, that is, in hygiene or in medicine; studies of digestion made on animals are evidently comparable with the same phenomena in man, as W. Beaumont's observations on his Canadian, compared with those he made by means of a gastric fistula in a dog, have superabundantly proved. Experiments made with animals, whether on the cerebrospinal nerves or on the vasomotor and secretory nerves of the large sympathetic (like experiments on circulation), are applicable at every point to the physiology and pathology of man. Experiments on animals, with deleterious substances or in harmful circumstances, are very useful and entirely conclusive for the toxicology and hygiene of man. Investigations of medicinal or of toxic substances also are wholly applicable to man from the therapeutic point of view; for, as I have shown,[18] the effects of these substances are the same on man as on animals, save for differences in degree. In pathological physiology, investigations of the formation of callus, the production of pus, etc., in animals are incontestably useful to human medicine.

But aside from all the connections to be found between man and animals we must recognize that there are differences also. Thus from a physiological point of view, experimental study of sense organs and cerebral functions must be made on man, on the one hand, because man is made higher than the animals by faculties which animals lack, and, on the other hand, because animals cannot directly account to us for the sensations which they experience. From the pathological point of view, we also note differences between man and animals; thus animals have parasitic and other diseases unknown to man, and *vice versa*. Among these diseases some are transmissible from man to animals and from animals to man; others, not. Finally, certain susceptibilities to inflammation of the peritoneum and other organs are not developed to the same degree in man as in animals of various classes or species. But, far from being motives to hold us back from experimenting and from applying conclusions from pathological investigation made on animals to differences observed in man, these differences provide convincing reasons to the contrary. Different species of animals show numerous and important differences in pathological tendencies. I have already said that there are breeds and varieties among domestic animals, such as asses, dogs and horses, which present wholly individual physiological and pathological susceptibilities; I have even noted individual differences that were often rather marked. Only experimental studies of these diversities can furnish an explanation of the individual differences observed in man, either in different races or in different in-

[18] Claude Bernard, *Recherches sur l'opium et ses alcaloïdes* (*Comptes rendus de l'Académie des sciences*, 1864).

dividuals of the same race, differences which physicians call predispositions or idiosyncrasies. Instead of persisting as indeterminate states of the organism, predispositions, when studied experimentally, will be classed in due time as particular cases of a general physiological law, which will thus become the scientific foundation of practical medicine.

To sum up, I not only conclude that experiments made on animals from the physiological, pathological and therapeutic points of view have results that are applicable to theoretic medicine, but I think that without such comparative study of animals, practical medicine can never acquire a scientific character. In this connection I shall finish with the words of Buffon, to which we might ascribe a different philosophic meaning, but which are scientifically very true for this occasion: "If animals did not exist, man's nature would be still more incomprehensible."

VIII. Comparison between Animals and Comparative Experimentation

In animals, and especially the higher animals, experimentation is so complex and liable to so many sources of error, both foreseen and unforeseen, that we must proceed most circumspectly to avoid them. To bring experimentation to bear on parts of the organism that we wish to explore, we must often do considerable tearing down and produce direct or indirect disturbances which must change or destroy our experimental results. These very real difficulties have often vitiated experimental investigations on living beings and furnished arguments to the detractors of experimentation. But science would never progress if we thought ourselves justified in renouncing scientific methods because they were imperfect; in this case, the one thing to do is to perfect the methods. Now perfecting physiological experimentation consists not only in improving instruments and operative methods, but above all and still more in study and well-regulated use of comparative experimentation.

We have elsewhere said (p. 55) that experimental counterproof must not be mistaken for comparative experimentation. Counterproof has not the slightest reference to sources of error that may be met in observing facts; it assumes that they are all avoided and is concerned only with experimental reasoning; it has in view only judging whether the relation established between a phenomenon and its immediate cause is correct and rational. Counterproof is therefore only a synthesis verifying an analysis or an analysis controlling a synthesis.

Comparative experimentation, on the contrary, bears solely on notation of fact and on the art of disengaging it from circumstances or from other phenomena with which it may be entangled. Comparative experimentation, however, is not exactly what philosophers call the method of differences. When an experimenter is confronted with complex phenomena due to the combined properties of various bodies, he proceeds by differentiation, that is to say, he separates each of these bodies, one by one in succession, and sees by the difference what part of the total phenomenon belongs to each of them. But this method of exploration implies two things: it implies, first of all, that we know how many bodies are concerned in expressing the whole phenomenon, and then it admits that these bodies do not combine in any such way as to confuse their action in a final harmonious result. In physiology the method of differences is rarely applicable, because we can never flatter ourselves that we know all the bodies and all the conditions combining to express a collection of phenomena, and in numberless cases because various organs of the body may take each other's place in phenomena, that are partly common to them all, and may more or less obscure the results of ablation of a limited part. Suppose, for instance, that we paralyze the whole body, a single muscle at a time. The disturbance produced by each paralyzed muscle will be more or less compensated and replaced by neighboring muscles, and we should finally come to the conclusion that each particular muscle contributed little to the movements of the body. The nature of this source of error has been very well expressed by comparing it with what would happen to an experimenter who removed, one after another, every brick in the foundation of a column. He would see, indeed, that removing in succession one brick at a time does not make the column totter, and he would come to the logical but false conclusion that not one of these bricks helps to support the column. In physiology, comparative experimentation depends upon quite another idea; for its object is to reduce the most complex investigation to unity, and its result is to eliminate by a single stroke all known and unknown sources of error.

Physiological phenomena are so complex that we could never experiment at all rigorously on living animals if we necessarily had to define all the other changes we might cause in the organism on which we were operating. But fortunately it is enough for us completely to isolate the one phenomenon on which our studies are brought to bear, separating it by means of comparative experimentation from all surrounding complications. Comparative experimentation reaches this goal by adding to a similar organism, used for comparison, all our experimental changes save one, the very one which we intend to disengage.

If, for instance, we wish to know the result of section or ablation of a deep-seated organ which cannot be reached without injuring many neighboring organs, we necessarily risk confusion in the total result between the effects of lesions caused by our operative procedure and the particular effects of section or ablation of the organ whose physiological rôle we wish to decide. The only way to avoid this mistake is to perform the same operation on a similar animal, but without making the section or ablation of the organ on which we are experimenting. We thus have two animals in which all the experimental conditions are the same, save one,—ablation of an organ whose action is thus disengaged and expressed in the difference observed between the two animals. Comparative experimentation in experimental medicine is an absolute and general rule applicable to all kinds of investigation, whether we wish to learn the effects of various agents influencing the bodily economy or to verify the physiological rôle of various parts of the body by experiments in vivisection.

At times comparative experimentation may be done on two animals of the same species in condition as closely comparable as possible; again the experiment must be made on the same animal. When working on two animals, as we have just said, we must place them in the same conditions, save one, the one that we wish to compare. This implies that the two animals compared are so much alike that differences noted in them after the experiment cannot be attributed to a difference depending on the individuals themselves. For experimenting on organs or tissues whose properties are definite and easily perceived, comparison of two animals of the same species will suffice; but, on the other hand, when we wish to compare delicate and fugitive qualities, we must make our comparison on the same animal, whether the nature of the experiment permits experimenting on him repeatedly at different times, or whether we have to act at one and the same time on similar parts of the same specimen. Differences, in fact, are harder to grasp in proportion as the phenomena that we wish to study grow more fugitive or more delicate; in this respect no animal is ever absolutely comparable with another, and, as we have already said, neither is the same animal comparable with himself at different times when we examine him, whether because he is in different conditions, or because his organism has grown less sensitive, by getting used to the substance given him or to the operation to which he is subjected.

IX. THE USE OF CALCULATION IN STUDY OF LIVING BEINGS; AVERAGES AND STATISTICS

Finally, it sometimes becomes necessary to extend comparative experimentation outside of the animal, since sources of error may also be met in the instruments used for experimentation. I shall limit myself here to pointing out and defining the principle of comparative experimentation; it will be explained, *à propos* of special cases, in the course of this work. In the third part of this Introduction I shall cite examples chosen to show the importance of comparative experimentation, which is the true foundation of experimental medicine; it would be easy, in fact, to prove that almost all experimental errors come from neglecting comparative judgment of facts or from thinking cases comparable which are not so.

In the experimental sciences, measurement of phenomena is fundamental, since their law can be established by quantitatively determining an effect in relation to a given cause. In biology, if we wish to learn the laws of life, we must therefore not only observe and note vital phenomena, but moreover must also define numerically the ratios of their relative intensity one to another.

The application of mathematics to natural phenomena is the aim of all science, because phenomenal law should always be mathematically expressed. To this end, data used in calculations should be results of well-analyzed facts, so that we may be sure that we fully know the conditions of the phenomena between which we wish to establish an equation. Now, I think that efforts of this kind are premature in most vital phenomena, precisely because these phenomena are so complex that we must not only assume, but are in fact certain that, beside the few among their conditions which we know, there are numberless others which are still totally unknown. I believe that the most useful path for physiology and medicine to follow now is to seek to discover new facts instead of trying to reduce to equations the facts which science already possesses. This does not mean that I condemn the application of mathematics to biological phenomena, because the science will later be established by this alone; only I am convinced that, since a complete equation is impossible for the moment, qualitative must necessarily precede quantitative study of phenomena.

Physicists and chemists have already often tried to reduce the physico-chemical phenomena of living beings to figures. Among the ancients, as well as among the moderns, the most eminent physicists and chemists wished to establish principles of animal mechanics and laws for chemical statistics of animals. Though the progress of physico-chemical science has made these problems more accessible to-day than in the past, it seems to me impossible to reach accurate conclusions at present, because foundations are lacking on which to base our calculations. We may, of course, strike a balance between what a living organism takes in as nourishment and what it gives out in excretions; but the results would be mere statistics incapable of throwing light on the inmost phenomena of nutrition in living beings. According to a Dutch chemist's phrase, this would be like trying to tell what happens inside a house by watching what goes in by the door and what comes out by the chimney. We can accurately fix the extreme terms of nutrition; but if we afterward try to interpret the intermediary between them, we find ourselves in an unknown region the greater part of which is created by the imagination, and this the more easily because figures often lend themselves marvellously to demonstrating the most diverse hypotheses. Twenty-five years ago, at the outset of my career as a physiologist, I was one of the first, I think, to carry experimentation into the inner environment of the organism, so as to follow experimentally, step by step, all the transformations of substances that chemists explained theoretically. I therefore devised experiments to investigate how sugar, one of the best defined of alimentary substances, is broken down in living beings. But instead of informing myself about the breaking down of sugar, my experiments led me to discover [19] that sugar is continually produced in animals, no matter what they eat. Moreover, these in-

[19] See the third part of this Introduction.

vestigations convinced me that numberless very complex physico-chemical phenomena take place, in the organic environment, which give rise to many other products, still unknown to us, with which the chemists do not at all reckon in their static equations. In the chemical statics of life, as well as in the various quantitative esti-mates of physiological phenomena, certainly neither chemical think-ing nor rigor in calculation is lacking; but physiological foundations, which most of the time are false, simply because they are incomplete. We are afterwards led astray all the more easily because we start from an incomplete experimental result and reason without verify-ing our deductions at every step. Let me cite examples of calcula-tions which I condemn, taking them from works which I neverthe-less hold in the highest esteem. In 1852 Bidder and Schmidt of Dorpat published highly important works on digestion and nutri-tion. Their investigations include excellent and very numerous raw data, but in my opinion the deductions from their calculations are often risky or erroneous. Thus, for example, they took a dog weigh-ing 16 kilograms; in the duct of the submaxillary gland they placed a tube through which the secretion flowed; and in one hour they ob-tained 5.640 grams of saliva, from which they concluded that for both glands this should make 11.280 grams. They afterward placed another tube in the duct of the same animal's parotid gland; and in an hour they obtained 8.790 grams of saliva which for both parotid glands would make 17.580 grams. Now, they went on, if we wish to apply these numbers to man, we must take a man weighing 64 kilograms or about four times as much as the dog in question; a calculation based on this ratio consequently gives us, for the man's submaxillary glands, 46 grams of saliva per hour, or 1.082 kilo-grams per day. For the parotid glands, we have 70 grams per hour, or 1.687 kilograms per day which reduced one half gives about 1.40 kilograms of saliva secreted in twenty-four hours by the salivary glands of an adult man.[20]

As the authors themselves feel, only one thing is true in the above: the crude result found in the dog; all the calculations deduced from this rest on false or doubtful foundations; first of all, doubling the product of one gland to get the product of both is incorrect, because physiology teaches us that in most cases double glands secrete alter-

[20] Bidder and Schmidt, *Die Verdauungssäfte und der Stoffwechsel.* Mittau and Leipzig, 1852. p. 12.

nately, and that, when one secretes a great deal, the other secretes less; then, besides the two submaxillary and parotid salivary glands, there are others which are not mentioned. Next, it is a mistake to believe that multiplying one hour's output of saliva by 24 gives the saliva poured into an animal's mouth in 24 hours. In fact, salivary secretion is highly intermittent and takes place only at meal time or when stimulated; during the rest of the time, the secretion is nil or insignificant. Finally, the quantity of saliva got from the salivary glands of the dog in this experiment was not absolute; it would have been nil if the mucous membrane of the mouth had not been stimulated; it might have been greater or less if another stimulant, stronger or weaker than vinegar, had been used.

Now the application of the above calculations to man is still more questionable. If the quantity of saliva had been multiplied by the weight of the salivary glands, a closer relation would have been found; but I cannot concede the validity of calculating the quantity of saliva from the weight of the body taken as a whole. Estimating a phenomenon in kilograms of the animal's body seems to me wholly incorrect, when all sorts of tissues foreign to the phenomenon in question are included.

In the part of their investigation devoted to nutrition, Bidder and Schmidt described a very notable experiment, perhaps one of the most laborious ever performed. From the point of view of elementary analysis, they kept a balance sheet of everything taken in and given out by a cat during eight days' nourishment and nineteen days' fasting. But this cat was in a physiological condition of which they were unaware; she was pregnant, and she had her kittens on the seventeenth day of the experiment. In these circumstances, our authors considered the kittens as excretions, and calculated them with other eliminated materials as a simple loss of weight.[21] I believe that these interpretations should be rectified when trying to define such complex phenomena.

In a word, I think that, if figures correspond with reality in these works of chemical statics applied to vital phenomena, it is only by chance or because the experimenters' feeling guides and rearranges the calculation. I repeat, nevertheless, that the criticism which I have just made is not directed against the principle of using calculations in physiology, but against its application under present

[21] Bidder and Schmidt, loc. cit., p. 397.

conditions. I am fortunate, moreover, in being able here to rely on the opinion of the physicists and chemists most competent in such matters. Regnault and Reiset, in their fine work on respiration, express themselves as follows about the calculations used to establish the theory of animal heat: "We have no doubt that animal heat is produced wholly by chemical reactions occurring in the bodily economy; but we think the phenomenon much too complex for possible calculation of the heat from the quantity of oxygen consumed. The substances burned in respiration are generally composed of carbon, hydrogen, nitrogen or oxygen, often in considerable proportions; when they are completely destroyed in respiration, the oxygen which they contain contributes to the formation of water and carbonic acid; and the heat liberated is therefore necessarily quite different from what would be produced in burning the supposedly free carbon and hydrogen. These substances, moreover, are not wholly destroyed; a portion is transformed into other substances which play special parts in the animal economy or escape, in excretions, in the form of highly oxidized materials (urea, uric acid). Now, in all these transformations and in the assimilation of substances taking place in the organs, heat is liberated or absorbed; but the phenomena are obviously so complex that there is little chance that we shall ever succeed in reducing them to calculation. It was therefore by a fortuitous circumstance in the experiments of Lavoisier, Dulong and Despretz, that the quantity of heat liberated by an animal was found to be about equal to what the carbon (contained in the carbonic acid produced) and the hydrogen would give off in burning,—the quantity of hydrogen being determined by a quite gratuitous assumption that the quantity of oxygen consumed, but not found in the carbonic acid, had been used in turning the hydrogen into water." [22]

Chemico-physical phenomena of living organisms are therefore still too complex to-day to be embraced as a whole, except by means of hypotheses. To find correct solutions of such vast problems, we must begin by analyzing the results of complicated reactions, and by separating them experimentally into distinct and simple questions. In several attempts which I have made on this analytic path, I have shown that we should not handle the problem of nutrition *en bloc*,

[22] Cf. Regnault and Reiset, *Recherches chimiques sur la respiration des animaux des diverses classes.* (*Annales de chimie et de physique.* 3d Series, Vol. XXVI, p. 217.)

but rather should first define the nature of the physico-chemical phenomena taking place in an organ made of some definite tissue, such as a muscle, gland or nerve; that we must at the same time take account of the organ's state of activity or rest. I have also shown that we can regulate an organ's state of rest or activity at will, by means of its nerves, and that we can even act on it locally without reverberation through the organism, if we first separate the peripheral nerves from the nervous centres.[23] When we have analyzed the physico-chemical phenomena peculiar to each tissue and each organ, then only can we try to understand nutrition as a whole and to found biochemistry on a solid base, that is to say, on the study of definite, complete and comparable physiological facts.

Another very frequent application of mathematics to biology is the use of averages which, in medicine and physiology, leads, so to speak, necessarily to error. There are doubtless several reasons for this; but the greatest obstacle to applying calculation to physiological phenomena is still, at bottom, the excessive complexity which prevents their being definite and comparable one with another. By destroying the biological character of phenomena, the use of *averages* in physiology and medicine usually gives only apparent accuracy to the results. From our point of view, we may distinguish between several kinds of averages: physical averages, chemical averages and physiological and pathological averages. If, for instance, we observe the number of pulsations and the degree of blood pressure by means of the oscillations of a manometer throughout one day, and if we take the average of all our figures to get the true or average blood pressure and to learn the true or average number of pulsations, we shall simply have wrong numbers. In fact, the pulse decreases in number and intensity when we are fasting and increases during digestion or under different influences of movement and rest; all the biological characteristics of the phenomenon disappear in the average. Chemical averages are also often used. If we collect a man's urine during twenty-four hours and mix all this urine to analyze the average, we get an analysis of a urine which simply does not exist; for urine, when fasting, is different from urine during digestion. A startling instance of this kind was invented by a physiologist

---

[23] Claude Bernard, *Sur le changement de couleur du sang dans l'état de fonction et de repos des glandes.—Analyse du sang des muscles au repos et en contraction. Leçons sur les liquides de l'organisme.* Paris, 1859.

who took urine from a railroad station urinal where people of all nations passed, and who believed he could thus present an analysis of *average* European urine! Aside from physical and chemical, there are physiological averages, or what we might call average descriptions of phenomena, which are even more false. Let me assume that a physician collects a great many individual observations of a disease and that he makes an average description of symptoms observed in the individual cases; he will thus have a description that will never be matched in nature. So in physiology, we must never make average descriptions of experiments, because the true relations of phenomena disappear in the average; when dealing with complex and variable experiments, we must study their various circumstances, and then present our most perfect experiment as a type, which, however, still stands for true facts. In the cases just considered, averages must therefore be rejected, because they confuse, while aiming to unify, and distort while aiming to simplify. Averages are applicable only to reducing very slightly varying numerical data about clearly defined and *absolutely simple* cases.

Let me further point out that the reduction of physiological phenomena to an expression in kilograms of body weight is vitiated by many sources of errors. For a certain number of years this method has been used by physiologists studying the phenomena of digestion (see p. 131). We observe, for instance, how much oxygen or how much food an animal consumes in a day; we then divide by the animal's weight and get the intake of food or of oxygen per kilogram. This method may also be applied to measure the action of toxic or medicinal materials. We poison an animal with a maximum dose of strychnine or curare, and divide the amount by the weight of the body, to get the amount of poison per kilogram. For greater accuracy in the experiments just cited, we should have to calculate, not per kilogram of the animal's body taken as a whole, but per kilogram of blood and of the unit on which the poison acts; otherwise we could not deduce any direct law from the reductions. But other conditions would still remain to be established similarly by experiment, conditions varying with age, height, state of digestion, etc.; in these measures, physiological conditions should always hold first rank.

To sum up, every possible application of calculation would be excellent if the physiological conditions were quite accurately defined. Physiologists and physicians should therefore concentrate their effort, for the moment, on defining these conditions. We must first accurately define the conditions of each phenomenon; this is true biological accuracy, and, without this preliminary study, all numerical data are inaccurate, and the more inaccurate because they include figures which mislead and impose on us by a false appearance of accuracy.

As for statistics, they are given a great rôle in medicine, and they therefore raise a medical question which we should examine here. The first requirement in using statistics is that the facts treated shall be reduced to comparable units. Now this is very often not the case in medicine. Everyone familiar with hospitals knows what errors may mark the definitions on which statistics are based. The names of diseases are very often given haphazard, either because the diagnosis is obscure, or because the cause of death is carelessly recorded by a student who has not seen the patient, or by an employee unfamiliar with medicine. For this reason pathological statistics can be valid only when compiled from data collected by the statistician himself. But even then, no two patients are ever exactly alike; their age, sex, temperament and any number of other circumstances involve differences, with the result that the average, or the relation deduced from our comparison of facts, may always be contested. But I cannot accept even the hypothesis that facts can ever be absolutely alike and comparable in statistics; they must necessarily differ at some point, for statistics would otherwise lead to absolute scientific results, while they can actually show only probability, never certainty. I acknowledge my inability to understand why results taken from statistics are called *laws;* for in my opinion scientific law can be based only on certainty, on absolute determinism, not on probability. I should stray from my subject, if I went into all possible explanation of the value of statistical methods based on the calculus of probabilities; yet I cannot but say here what I think about the application of statistics to physiological science in general and to medicine in particular.

In every science, we must recognize two classes of phenomena, first, those whose cause is already defined; next, those whose cause is still undefined. With phenomena whose cause is defined, statistics have nothing to do; they would even be absurd. As soon as the circumstances of an experiment are well known, we stop gathering statistics: we should not gather cases to learn how often water is made of oxygen and hydrogen; or when cutting the sciatic nerve, to learn how often the muscles to which it leads will be paralyzed. The effect will occur always without exception, because the cause of the phenomena is accurately defined. Only when a phenomenon includes conditions as yet undefined, can we compile statistics; we must learn, therefore, that we compile statistics only when we cannot possibly help it; for in my opinion statistics can never yield scientific truth, and therefore cannot establish any final scientific method. A single example will illustrate my meaning. Certain experimenters, as we shall later see, published experiments by which they found that the anterior spinal roots are insensitive; other experimenters published experiments by which they found that the same roots were sensitive. These cases seemed as comparable as possible; here was the same operation done by the same method on the same spinal roots. Should we therefore have counted the positive and negative cases and said: the law is that anterior roots are sensitive, for instance, 25 times out of a 100? Or should we have admitted, according to the theory called the law of large numbers, that in an immense number of experiments we should find the roots equally often sensitive and insensitive? Such statistics would be ridiculous, for there is a reason for the roots being insensitive and another reason for their being sensitive; this reason had to be defined; I looked for it, and I found it; so that we can now say: the spinal roots are always sensitive in given conditions, and always insensitive in other equally definite conditions.

I will cite still another example borrowed from surgery. A great surgeon performs operations for stone by a single method; later he makes a statistical summary of deaths and recoveries, and he concludes from these statistics that the mortality law for this operation is two out of five. Well, I say that this ratio means literally nothing scientifically and gives us no certainty in performing the next operation; for we do not know whether the next case will be among the recoveries or the deaths. What really should be done, instead of gathering facts empirically, is to study them more accurately, each in its special determinism. We must study cases of death with great care and try to discover in them the cause of mortal accidents, so as to master the cause and avoid the accidents. Thus, if we accurately know the cause of recovery and the cause of death, we shall always have a recovery in a definite case. We cannot, indeed, admit that cases with different endings were identical at every point. In the patient who succumbed, the cause of death was evidently something which was not found in the patient who recovered; this something we must determine, and then we can act on the phenomena or recognize and foresee them accurately. But not by statistics shall we succeed in this; never have statistics taught anything, and never can they teach anything about the nature of phenomena. I shall further apply what I have just said to all the statistics compiled with the object of learning the efficacy of certain remedies in curing diseases. Aside from our inability to enumerate the sick who recover of themselves in spite of a remedy, statistics teach absolutely nothing about the mode of action of medicine nor the mechanics of cure in those in whom the remedy may have taken effect.

It is said that coincidence may play so large a part in causes of statistical errors, that we should base conclusions only on large numbers. But physicians have nothing to do with what is called the *law of large numbers,* a law which, according to a great mathematician's expression, is always true in general and false in particular. This amounts to saying that the law of large numbers never teaches us anything about any particular case. What a physician needs to know is whether his patient will recover, and only the search for scientific determinism can lead to this knowledge. I do not understand how we can teach practical and exact science on the basis of statistics. The results of statistics, even statistics of large numbers, seem indeed to show that some compensation in the variations of phenomena leads to a law; but as this compensation is indefinite, even the mathematicians confess that it can never teach us anything about any particular case; for they admit that if the red ball comes out fifty times in succession, that is no reason why a white ball would be more likely to come out the fifty-first time.

Statistics can therefore bring to birth only conjectural sciences; they can never produce active experimental sciences, i.e., sciences which regulate phenomena according to definite laws. By statistics, we get a conjecture of greater or less probability about a given case, but never any certainty, never any absolute determinism. Of course, statistics may guide a physician's prognosis; to that extent they are useful. I do not therefore reject the use of statistics in medicine, but I condemn not trying to get beyond them and believing in statistics as the foundation of medical science. This false idea leads certain physicians to believe that medicine cannot but be conjectural; and from this, they infer that physicians are artists who must make up for the indeterminism of particular cases by medical tact. Against these anti-scientific ideas we must protest with all our power, because they help to hold medicine back in the lowly state in which it has been so long. All sciences necessarily began by being conjectural; even to-day science has its conjectural parts. Medicine is still almost wholly conjectural. I do not deny it; I only mean to say that modern medical science must exert itself to get out of the temporary condition which is no more a final scientific state for medicine than for any other science. The scientific state will be harder to reach and will take longer to establish in medicine, because of the complexity of the phenomena; but the goal of scientific physicians in their own science, as in the rest, is to reduce the indeterminate to the determinate. Statistics therefore apply only to cases in which the cause of the facts observed is still indeterminate. In these circumstances, statistics in my opinion can serve only to guide the observer toward investigation of the indeterminate cause, but they can never lead to any real law. I emphasize this point, because many physicians have great confidence in statistics when based on well-observed facts which they consider mutually comparable, and they believe that such statistics may lead to knowledge of phenomenal law. I have already said that facts are never identical; therefore statistics are simply an empirical enumeration of observations.

In a word, if based on statistics, medicine can never be anything but a conjectural science; only by basing itself on experimental determinism can it become a true science, i.e., a sure science. I think of this idea as the pivot of experimental medicine, and in this respect experimental physicians take a wholly different point of view from so-called observing physicians. Indeed, if a phenomenon appears just once in a certain aspect, we are justified in holding that, in the same conditions, it must always appear in the same way. If, then, it differs in behavior, the conditions must be different. But indeterminism knows no laws; laws exist only in experimental determinism, and without laws there can be no science. Most physicians seem to believe that, in medicine, laws are elastic and indefinite. These are false ideas which must disappear if we mean to found a scientific medicine. As a science, medicine necessarily has definite and precise laws which, like those of all the sciences, are derived from the criterion of experiment. To the explanation of these ideas I shall especially devote the work which I have named *Principles of Experimental Medicine,* in order to show that the principles of experimental determinism must be applied to medicine, if it is to become an exact science founded on experimental determinism, instead of remaining a conjectural science based on statistics. A conjectural science may indeed rest on the indeterminate; but an experimental science accepts only determinate or determinable phenomena.

Only determinism in an experiment yields absolute law; and he who knows the true law is no longer free to see a phenomenon otherwise. The indeterminism of statistics leaves to thought a certain liberty limited by the numbers themselves; and in this sense philosophers were able to say that liberty begins where determinism ends. But when determinism increases, statistics can no longer grasp and confine it within a limit of variations. There we leave science, for we are forced to invoke chance or an occult cause to regulate phenomena. We shall certainly never reach absolute determinism in everything; man could no longer exist. There will always be some indeterminism then, in all the sciences, and more in medicine than in any other. But man's intellectual conquest consists in lessening and driving back indeterminism in proportion as he gains ground for determinism by the help of the experimental method. This alone should satisfy his ambition, for by this alone is he extending, and can he further extend, his power over nature.

X. The Physiologist's Laboratory and Various Methods Necessary to the Study of Experimental Medicine

Every experimental science requires a laboratory. There the man of science withdraws, and by means of experimental analysis tries to understand phenomena that he has observed in nature.

A physician's subject of study is necessarily the patient, and his first field for observation is the hospital. But if clinical observation teaches him to know the form and course of diseases, it cannot suffice to make him understand their nature; to this end he must penetrate into the body to find which of the internal parts are injured in their functions. That is why dissection of cadavers and microscopic study of diseases were soon added to clinical observation. But to-day these various methods no longer suffice; we must push investigation further and, in analyzing the elementary phenomena of organic bodies, must compare normal with abnormal states. We showed elsewhere how incapable is anatomy alone to take account of vital phenomena, and we saw that we must add study of all physico-chemical conditions which contribute necessary elements to normal or pathological manifestations of life. This simple suggestion already makes us feel that the laboratory of a physiologist-physician must be the most complicated of all laboratories, because he has to experiment with phenomena of life which are the most complex of all natural phenomena.

Libraries may also be considered as part of the laboratory of a man of science or experimenting physician. But this is on condition that he shall read the observations, experiments and theories of his predecessors in order to know them and verify them in nature, and not to find opinions ready-made in books, thus saving himself the trouble of working and of trying to further the investigation of natural phenomena. Misconceived erudition has been, and still is, one of the greatest obstacles to the advancement of experimental science. Thus erudition, setting man's authority in the place of facts, halted science through several centuries at Galen's ideas, without any one's daring to touch them; and this scientific superstition was such that Mundini and Vesalius, who first contradicted Galen by confronting his opinions with animal dissections, were considered innovators and revolutionaries. Yet such should always be the practice of scientific erudition. It should always be accompanied by critical investigations of nature, planned to verify the facts about which we speak, and to decide the opinions which we discuss. In this way science in advancing would be simplified and cleansed by sound experimental criticism, instead of being encumbered by exhuming an accumulation of numberless facts and opinions among which it is soon impossible to distinguish falsehood from truth. It would be out of place for me here to say more of the mistakes and misdirection of most of the studies of medical literature, characterized as historical or philosophical. I may perhaps have occasion to explain myself elsewhere on this subject; for the moment, I shall limit myself to saying that, in my opinion, all these mistakes have their origin in a perpetual confusion between literary or artistic production and scientific production, between criticism of art and scientific criticism, between the history of science and the history of men.

Literary and artistic productions never grow old, in this sense, that they are expressions of feeling, changeless as human nature. We may add that philosophical ideas stand for aspirations of the human spirit which are also of all time. But science, which stands for what man has learned, is essentially mobile in expression; it varies and perfects itself in proportion to the increase of acquired knowledge. Present day science is therefore necessarily higher than the science of the past; and there is no sort of reason for going in search of any addition to modern science through knowledge of the ancients. Their theories, necessarily false because they do not include facts discovered since then, can be of no real advantage to contemporary science. No experimental science, then, can make progress except by advancing and pursuing its work in the future. It would be absurd to believe that we should go in search of it in the study of books bequeathed to us by the past. We can find there only the history of the human mind, which is quite another matter.

We must of course be familiar with what we call scientific literature, and know what our predecessors have done. But scientific criticism in the literary manner can be of no possible use to science. Indeed, we need not ourselves be poets or artists to judge literary or artistic work, but this is not true of experimental science. We cannot judge of a memoir on chemistry without being chemists nor of a memoir on physiology if we are not physiologists. In deciding between two different scientific opinions, it is not enough to be a good philologist or a good translator, we must above all be deeply versed in technical science; we must even be masters of the special science and ourselves be able to experiment and do better than the men whose opinions we discuss. Some time ago I discussed an anatomical question concerning the anastomoses of the pneumogastric and spinal nerves.[24] Willis, Scarpa and Bischoff had expressed different and even opposite opinions on this subject. A mere scholar could only have quoted these various opinions and more or less correctly compared the texts; that would not have answered the scientific question. It was therefore necessary to dissect and

[24] Claude Bernard, *Recherches expérimentales sur les fonctions du nerf spinal.* (*Mémoires présentés par divers savants étrangers à l'Académie des sciences.* Vol. X. 1851.)

to perfect our methods of dissection, so as to follow the nervous anastomoses more precisely and to compare each anatomist's description with nature. This is what I did, and I found that the difference between the authors in question came from their not having assigned the same limits to the nerves. So anatomy, carried further, explained their anatomical dissension. I therefore refuse to acknowledge that science has a place for men who make criticism their specialty, as in letters and in the arts. To be really useful, criticism in every science must be done by men of science themselves, and by the most eminent masters.

Another somewhat frequent error consists in confusing the history of man with the history of some science. Theological and didactic evolution of experimental science is by no means expressed in the chronological history of the men concerned with it. We must nevertheless except the mathematical and astronomical sciences; but this cannot apply to the physico-chemical experimental sciences or to medicine in particular. Medicine was born of need, said Baglivi, that is to say, from the first time that anyone was ill, men went to his aid and sought to cure him. From its cradle, medicine has therefore been an applied science mixed with religion and with the feelings of sympathy that men experienced one for another. But did medicine as a science exist? Evidently not. Continuing through centuries as blind empiricism, it enriched itself, little by little and almost by chance, with observations and investigations in unrelated directions. Physiology, pathology and therapeutics developed as distinct sciences. That was the wrong road. Only to-day can we begin to see the conception of an experimental, scientific medicine in the fusion of these three in a single point of view.

The experimental point of view is a coronation of perfected science; for we must not deceive ourselves; true science exists only when man succeeds in accurately foreseeing the phenomena of nature and mastering them. Noting and classifying natural bodies and phenomena is not at all the equivalent of complete science. True science acts and explains its action or its power: that is its character, that is its aim. Let me amplify my idea. I have often heard physicians say that physiology or the explanation of vital phenomena in either the physiological or the pathological state is only a part of medicine, because medicine is knowledge of diseases in general. I have similarly heard zoölogists say that physiology, or the explanation of vital phenomena in all their variety, is only a dismemberment or specialty of zoölogy, because zoölogy is knowledge of animals in general. Talking in the same way, a geologist or a mineralogist might say that physics and chemistry are only dismemberments of geology or mineralogy, which include knowledge of the earth and of animals in general. Here are mistakes or at least misunderstandings which need to be explained. First of all, we must recognize that our divisions into sciences are not a part of nature; they exist only in the mind which, by reason of its infirmity, is forced to create categories of bodies and of phenomena, so as to understand them better by studying their characteristics or properties from special points of view. It follows that the same body may be studied mineralogically, physically, chemically, etc.; but in nature there is really neither chemistry nor physics, nor zoölogy, nor physiology, nor pathology; there are only bodies to be classified or phenomena to be known and mastered. Now the science that gives man means of analyzing and experimentally mastering phenomena is the furthest advanced. It must necessarily be the last established; that is no reason to consider it a dismemberment of earlier sciences. In this respect, physiology, which is the highest and most difficult science of living beings, cannot be regarded as a dismemberment of medicine or zoölogy, any more than physics or chemistry are dismemberments of geology or mineralogy. Physics and chemistry are the two active mineral sciences by means of which man can master the phenomena of inorganic bodies. Physiology is the vital, active science by whose aid man will be able to act on animals and on man, whether in health or in sickness. It would be a grave illusion for physicians to believe they know diseases by giving them names, because they classify and describe them, just as it would be an illusion for zoölogists or botanists to believe they know animals and vegetables because they have named them, catalogued, dissected and shut them up in museums, after stuffing, preparing or drying them. Physicians will not know diseases until they can act on them rationally and experimentally, just as zoölogists will not know animals until they explain and regulate the phenomena of life. To sum up, we must not be duped by our own works; we cannot assign any absolute value to scientific classifications, either in books or in academies. Men who leave the beaten track are innovators, and those who blindly persist in it hamper scientific progress. The very evolution of human knowledge means that experimental science must be the goal, and this evolution requires that earlier sciences of classification shall lose importance as the experimental sciences develop.

The spirit of man follows a necessary and logical course in the search for scientific truth. It observes facts, compares them, deduces appropriate results which it controls by experiment, to rise to more and more general propositions and truths. In this advancing labor, a man of science must, of course, know and deal with his predecessors' work. But he must be thoroughly convinced that this work is merely a support from which to go farther, and that new scientific truths are not to be found in study of the past, but rather in studies made anew on nature, i.e., in the laboratory. Useful scientific literature, then, is preëminently the scientific literature of modern work which enables us to keep up with scientific progress; and even this must not be carried too far, lest it dry up the mind and stifle invention and scientific originality. But what use can we find in exhuming worm-eaten theories or observations made without proper means of investigation? That may, of course, be helpful in learning the mistakes through which the human mind has passed in its evolution, but it is time wasted for science, properly speaking. I deem it highly important to guide the minds of students early toward active experimental science, by making them understand that it develops in laboratories, instead of leaving them to believe that it awaits them in books or in the interpretation of ancient writings. We know from history the sterility of the scholastic path and that science did not begin to soar until men substituted for the authority of books the authority of facts ascertained in nature with the help of more and more perfect experimental methods; Bacon's greatest merit was that he proclaimed this truth aloud. As for me, I think that turning medicine back to-day toward the belated and aged commentaries of antiquity is a retrogression, a return to scholasticism, while guiding medicine toward laboratories and toward experimental, analytical study of disease is an advance along the path of true progress, that is, toward the foundation of experimental medical science. With me, this is a deep conviction; I shall always seek to make it prevail both in my teaching and in my work.

A physiological laboratory, therefore, should now be the culminating goal of any scientific physician's studies; but here again I must explain myself to avoid misunderstanding. Hospitals, or rather hospital wards, are not physicians' laboratories, as is often believed; as we said before, these are only his fields for observation; there must be held what we call clinics, i.e., studies of disease as complete as possible. Medicine necessarily begins with clinics, since they determine and define the object of medicine, i.e., the medical problem; but while they are the physician's first study, clinics are not the foundation of scientific medicine; physiology is the foundation of scientific medicine because it must yield the explanation of morbid phenomena by showing their relations to the normal state. We shall never have a science of medicine as long as we separate the explanation of pathological from the explanation of normal, vital phenomena.

Here then lies the real medical problem; this is the foundation on which modern scientific medicine will be built. As we see, experimental medicine does not exclude clinical medicine; on the contrary, it comes only after it. But it is a higher science, and one necessarily more vast and general. We easily imagine how an observational or empirical physician, never leaving his hospital, may think medicine completely shut in there, as a science distinct from physiology, of which it feels no need. But for a man of science there is no separate science of medicine or physiology, there is only a science of life. There are only phenomena of life to be explained in the pathological as well as in the physiological state. By putting this fundamental idea and this general conception of medicine into the minds of young people at the outset of their medical studies, we shall show them that the physico-chemical sciences which they have learned are tools to help them analyze the phenomena of life in its normal and pathological states. In frequenting hospitals, amphitheatres and laboratories, they will easily grasp the general connection uniting all the medical sciences, instead of learning them like fragments of detached knowledge with no relation between them.

In a word, I consider hospitals only as the entrance to scientific medicine; they are the first field of observation which a physician enters; but the true sanctuary of medical science is a laboratory; only there can he seek explanations of life in the normal and pathological states by means of experimental analysis. I shall not concern myself here with the clinical side of medicine; I assume it as known or as still being perfected in hospitals by the new methods of diagnosis which physics and chemistry are constantly giving to symptomatology. In my opinion, medicine does not end in hospitals, as is often believed, but merely begins there. In leaving the hospital, a physician, jealous of the title in its scientific sense, must go into his laboratory; and there, by experiments on animals, he will seek to account for what he has observed in his patients, whether about the action of drugs or about the origin of morbid lesions in organs or tissues. There, in a word, he will achieve true medical science. Every scientific physician should, therefore, have a physiological laboratory; and this work is especially intended to give physicians rules and principles of experimentation to guide their study of experimental medicine, that is, their analytic and experimental study of disease. The principles of experimental medicine, then, will be simply the principles of experimental analysis applied to the phenomena of life in its healthy and its morbid states.

The biological sciences to-day are no longer seeking their path. Because of their complex nature they vacillated longer than other simpler sciences in the regions of philosophy and system; but they launched at last into the experimental path where to-day they are well advanced. So they now need only one thing more, and that is means of development. Such means are laboratories and all the conditions and instruments necessary to cultivate the scientific field of biology.

To the honor of French science, it must be stated that it had the glory of decisively inaugurating the experimental method in the science of vital phenomena. Toward the end of the last century, the renewal of chemistry strongly influenced the advance of physiological science, and the work of Lavoisier and Laplace on breathing cleared a fertile path for analytic physico-chemical experimentation on the phenomena of life. My teacher, Magendie, who was led into a medical career by this same influence, devoted his life to advocating experimentation in the study of physiological phenomena. Nevertheless, application of the experimental method to animals was hindered from the first by the lack of suitable laboratories and by all sorts of difficulties which are disappearing to-day, but from which I myself often suffered in my youth. The scientific impulse, started in France, spread through Europe, and little by little the analytic experimental method entered the realm of biological science as a general method of investigation. But this method was perfected more, and it brought forth more fruit in countries where conditions for its development were more favorable. Throughout Germany to-day there are laboratories, called physiological institutes, which are admirably endowed and organized for the experimental study of vital phenomena. They exist in Russia also, where new ones of gigantic size are being built. Scientific production is naturally in proportion to the means of cultivation which a science possesses; there is nothing astonishing, then, in the fact that Germany, where the means of cultivating the physiological sciences are most liberally installed, is distancing other countries in the quantity of its scientific production. The genius of man, of course, cannot abdicate its supremacy in science. In experimental science, however, a scientific man is the prisoner of his ideas, if he does not learn to question nature for himself, and if he does not possess suitable and necessary tools. We cannot imagine a physicist or a chemist without his laboratory. But as for the physician, we are not yet in the habit of believing that he needs a laboratory; we think that hospitals and books should suffice. That is a mistake; clinical information no more suffices for physicians than knowledge of minerals suffices for chemists or physicists. Physiological physicians must experimentally analyze the phenomena of living matter, as physicists and chemists experimentally analyze the phenomena of inorganic matter. A laboratory is therefore a condition *sine qua non* of the development of experimental medicine, as it was for all the other physico-chemical sciences. Without it, neither experimenters nor experimental science can exist.

I shall no longer dwell on so important a subject which cannot here be sufficiently worked out; let me end by saying that one truth is well established in modern science, namely, that scientific courses can only serve to introduce and to create a taste for the sciences. By pointing out, from a professional chair, the results as well as the methods of a science, a teacher may form the minds of his hearers and make them apt in learning and choosing their own direction; but he can never make them men of science. The laboratory is the real nursery of true experimental scientists, i.e., those who create the science that others afterward popularize. Now if we want much fruit, we must first care for our nurseries of fruit trees. The evidence of this truth will necessarily bring about general and deep reform in scientific teaching. For, I repeat, it is to-day everywhere recognized that pure science germinates and develops in laboratories, to spread out later and cover the world with useful applications. We must, therefore, first of all attend to the scientific source, since applied science necessarily proceeds from pure science.

Science and men of science are cosmopolitans, and it seems hardly important whether a scientific truth develops at any particular spot on the globe, as long as the general diffusion of science allows all men to share in it. However I cannot help praying that my country, the evident promoter and protector of scientific progress and the starting point of the brilliant era through which experimental physical science [25] is now passing, may have great, public, physiological laboratories as soon as possible, so as to make pleiads of physiologists and young experimenting physicians. Only laboratories can teach the difficulties of science to those who frequent them; they show that pure science has always been the source of all the riches acquired by man and of all his real conquests over the phenomena of nature. This is also excellent education for the young, because it makes them understand that the present, very brilliant applications of science are merely the blossoming of earlier labors, and that those who reap the benefits to-day owe a tribute of gratitude to their predecessors who painfully cultivated the tree of science, but never saw its fruits.

I cannot here treat all the conditions necessary to a good laboratory of physiology or experimental medicine. That would obviously amount to summarizing everything still to be explained in this work. I shall therefore limit myself to adding one word. I said above, that the laboratory of a physiological physician must be the most complex of all laboratories, because the experimental analyses to be made there are the most complex of all, requiring the help of all other sciences. The laboratory of a medical physiologist must be connected with a hospital so as to receive the various pathological specimens on which scientific investigation is brought to bear. It must next include healthy and diseased animals for the study of questions of normal and pathological physiology. But as vital phenomena, whether in the normal or in the pathological state, are

[25] In 1771 a course in experimental physiology was given by Professor A. Portal at the Collège de France; the experiments were reported by Monsieur Collomb who published them in letter form in 1771; they were republished in 1808 with a few additions in the work by Portal entitled: *Mémoires sur la nature et le traitement de plusieurs maladies, avec le précis d'expériences sur les animaux vivants*. Paris, 1800-1825.

analyzed mainly by means of tools borrowed from physico-chemical science, instruments must necessarily be somewhat liberally provided. The solution of certain scientific questions often imperatively demands costly and complicated instruments, so that we may then say that scientific questions are secondary to the question of money. However, I do not approve the luxury as to instruments to which certain physiologists have yielded. In my opinion, we should seek to simplify instruments as much as possible, not only for pecuniary, but also for scientific reasons; for we need to learn that the more complicated the instrument, the more sources of error does it create. Experimenters do not grow great by the number and complexity of instruments; it is really the other way. The great experimenters, Berzelius and Spallanzani, made great discoveries by means of simple instruments. In the course of this work, our principle, then, will be to seek, as far as possible, to simplify means of study; for instruments must be allies, not sources of error because of their complications.

PART THREE

APPLICATIONS OF THE EXPERIMENTAL METHOD TO THE STUDY OF VITAL PHENOMENA

CHAPTER I

EXAMPLES OF EXPERIMENTAL PHYSIOLOGICAL INVESTIGATION

THE ideas explained in the first two parts of this introduction will be all the better understood if we can connect them with actual investigations in experimental physiology and medicine. For this reason, I have put together in the following part a certain number of examples that seem to me appropriate. As far as possible, I have quoted from myself in all these examples, for the sole reason that, in the matter of reasoning and intellectual processes, I shall be much more certain of what I describe in telling what has happened to me than in interpreting what may have taken place in the minds of others. I am not, however, so fatuous as to give these examples as models to follow; I use them only to express my ideas better and to make my thought easier to grasp.

In scientific investigations, various circumstances may serve as starting points for research; I will reduce all these varieties, however, to two chief types:

1. Where the starting point for experimental research is an observation;

2. Where the starting point for experimental research is an hypothesis or a theory.

I. WHERE THE STARTING POINT FOR EXPERIMENTAL RESEARCH IS AN OBSERVATION

Experimental ideas are often born by chance, with the help of some casual observation. Nothing is more common; and this is really the simplest way of beginning a piece of scientific work. We take a walk, so to speak, in the realm of science, and we pursue what happens to present itself to our eyes. Bacon compares scientific investigation with hunting; the observations that present themselves are the game. Keeping the same simile, we may add that, if the game presents itself when we are looking for it, it may also present itself when we are not looking for it, or when we are looking for game of another kind. I shall cite an example in which these two cases presented themselves in succession. At the same time I shall be careful to analyze every circumstance involved, so as to show how the principles apply which we explained in the first part of the introduction and especially in Chapters I and II.

First example.—One day, rabbits from the market were brought into my laboratory. They were put on the table where they urinated, and I happened to observe that their urine was clear and acid. This fact struck me, because rabbits, which are herbivora, generally have turbid and alkaline urine; while on the other hand carnivora, as we know, have clear and acid urine. This observation of acidity in the rabbits' urine gave me an idea that these animals must be in the nutritional condition of carnivora. I assumed that they had probably not eaten for a long time, and that they had been transformed by fasting, into veritable carnivorous animals, living on their own blood. Nothing was easier than to verify this preconceived idea or hypothesis by experiment. I gave the rabbits grass to eat; and a few hours later, their urine became turbid and alkaline. I then subjected them to fasting and after twenty-four hours or thirty-six hours at most, their urine again became clear and strongly acid; then after eating grass, their urine became alkaline again, etc. I repeated this very simple experiment a great many times, and always with the same result. I then repeated it on a horse, an herbivorous animal which also has turbid and alkaline urine. I found that fasting, as in rabbits, produced prompt acidity of the urine, with such an increase in urea, that it spontaneously crystallizes at times in the cooled urine. As a result of my experiments, I thus reached the general proposition which then was still unknown, to wit, that all fasting animals feed on meat, so that herbivora then have urine like that of carnivora.

We are here dealing with a very simple, particular fact which allows us easily to follow the evolution of experimental reasoning. When we see a phenomenon which we are not in the habit of seeing, we must always ask ourselves what it is connected with, or putting it differently, what is its proximate cause; the answer or the idea, which presents itself to the mind, must then be submitted to experiment. When I saw the rabbits' acid urine, I instinctively asked myself what could be its cause. The experimental idea consisted in the connection, which my mind spontaneously made, between acidity of the rabbits' urine, and the state of fasting which I considered equivalent to a true flesh-eater's diet. The inductive reasoning which I implicitly went through was the following syllogism: the urine of carnivora is acid; now the rabbits before me have acid urine, therefore they are carnivora, i.e., fasting. This remained to be established by experiment.

But to prove that my fasting rabbits were really carnivorous, a counterproof was required. A carnivorous rabbit had to be experimentally produced by feeding it with meat, so as to see if its urine would then be clear, as it was during fasting. So I had rabbits fed on cold boiled beef (which they eat very nicely when they are given nothing else). My expectation was again verified, and, as long as the animal diet was continued, the rabbits kept their clear and acid urine.

To complete my experiment, I made an autopsy on my animals, to see if meat was digested in the same way in rabbits as in carnivora. I found, in fact, all the phenomena of an excellent digestion in their intestinal reactions, and I noted that all the chyliferous vessels were gorged with very abundant white, milky chyle, just as in carnivora. But à propos of these autopsies which confirmed my ideas on meat digestion in rabbits, lo and behold a fact presented itself which I had not remotely thought of, but which became, as we shall see, my starting point in a new piece of work.

Second example.—(Sequel to the last)—In sacrificing the rabbits which I had fed on the meat, I happened to notice that the white and milky lymphatics were first visible in the small intestine at the lower part of the duodenum, about thirty centimeters below the pylorus. This fact caught my attention because in dogs they are first visible much higher in the duodenum just below the pylorus. On examining more closely, I noted that this peculiarity in rabbits coincided with the position of the pancreatic duct which was inserted very low and near the exact place where the lymphatics began to contain a chyle made white and milky by emulsion of fatty nutritive materials.

Chance observation of this fact evoked the idea which brought to birth the thought in my mind, that pancreatic juice might well cause the emulsion of fatty materials and consequently their absorption by the lymphatic vessels. Instinctively again, I made the following syllogism: the white chyle is due to emulsion of the fat; now in rabbits white chyle is formed at the level where pancreatic juice is poured into the intestine; therefore it is pancreatic juice that makes the emulsion of fat and forms the white chyle. This had to be decided by experiment.

In view of this preconceived idea I imagined and at once performed a suitable experiment to verify the truth or falsity of my suppositions. The experiment consisted in trying the properties of pancreatic juice directly on neutral fats. But pancreatic juice does not spontaneously flow outside of the body, like saliva, for instance, or urine; its secretory organ is, on the contrary, lodged deep in the abdominal cavity. I was therefore forced to use the method of experimentation to secure the pancreatic fluid from living animals in suitable physiological conditions and in sufficient quantity. Only then could I carry out my experiment, that is to say, control my preconceived idea; and the experiment proved that my idea was correct. In fact pancreatic juice obtained in suitable conditions from dogs, rabbits and various other animals, and mixed with oil or melted fat, always instantly emulsified, and later split these fatty bodies into fatty acids, glycerine, etc., etc., by means of a specific ferment.

I shall not follow these experiments further, having explained them at length in a special work.[1] I wish here to show merely how an accidental first observation of the acidity of rabbits' urine suggested to me the idea of making experiments on them with carnivorous feeding, and how later, in continuing these experiments, I brought to light, without seeing it, another observation concerning the peculiar arrangement of the junction of the pancreatic duct in rabbits. This second observation gave me, in turn, the idea of experimenting on the behavior of pancreatic juice.

From the above examples we see how chance observation of a fact or phenomenon brings to birth, by anticipation, a preconceived idea or hypothesis about the probable cause of the phenomenon ob-

---

[1] Claude Bernard, *Mémoire sur le pancréas et sur le rôle du suc pancréatique dans les phénomènes digestifs.* Paris, 1856.

served; how the preconceived idea begets reasoning which results in the experiment which verifies it; how, in one case, we had to have recourse to experimentation, i.e., to the use of more or less complicated operative processes, etc., to work out the verification. In the last example, experiment played a double rôle; it first judged and confirmed the provisions of the reasoning which it had begotten; but what is more, it produced a fresh observation. We may therefore call this observation an observation produced or begotten by experiment. This proves that, as we said, all the results of an experiment must be observed, both those connected with the preconceived idea and those without any relation to it. If we saw only facts connected with our preconceived idea, we should often cut ourselves off from making discoveries. For it often happens that an unsuccessful experiment may produce an excellent observation, as the following example will prove.

Third example.—In 1857, I undertook a series of experiments on the elimination of substances in the urine, and this time the results of the experiment, unlike the previous examples, did not confirm my previsions or preconceived ideas. I had therefore made what we habitually call an unsuccessful experiment. But we have already posited the principle that there are no unsuccessful experiments; for, when they do not serve the investigation for which they were devised, we must still profit by observation to find occasion for other experiments.

In investigating how the blood, leaving the kidney, eliminated substances that I had injected, I chanced to observe that the blood in the renal vein was crimson, while the blood in the neighboring veins was dark like ordinary venous blood. This unexpected peculiarity struck me, and I thus made observation of a fresh fact begotten by the experiment, but foreign to the experimental aim pursued at the moment. I therefore gave up my unverified original idea, and directed my attention to the singular coloring of the venous renal blood; and when I had noted it well and assured myself that there was no source of error in my observation, I naturally asked myself what could be its cause. As I examined the urine flowing through the urethra and reflected about it, it occurred to me that the red coloring of the venous blood might well be connected with the secreting or active state of the kidney. On this hypothesis, if the renal secretion was stopped, the venous blood should become dark: that is what happened; when the renal secretion was reëstablished, the venous blood should become crimson again; this I also succeeded in verifying whenever I excited the secretion of urine. I thus secured experimental proof that there is a connection between the secretion of urine and the coloring of blood in the renal vein.

But that is still by no means all. In the normal state, venous blood in the kidney is almost constantly crimson, because the urinary organ secretes almost continuously, though alternately for each kidney. Now I wished to know whether the crimson color is a general fact characteristic of the other glands, and in this way to get a clear-cut counterproof demonstrating that the phenomenon of secretion itself was what led to the alteration in the color of the venous blood. I reasoned thus: if, said I, secretion, as it seems to be, causes the crimson color of glandular venous blood, then, in such glandular organs as the salivary glands which secrete intermittently, the venous blood will change color intermittently and become dark, while the gland is at rest, and red during secretion. So I uncovered a dog's submaxillary gland, its ducts, its nerves and its vessels. In its normal state, this gland supplies an intermittent secretion which we can excite or stop at pleasure. Now while the gland was at rest, and nothing flowed through the salivary duct, I clearly noted that the venous blood was, indeed, dark, while, as soon as secretion appeared, the blood became crimson, to resume its dark color when the secretion stopped; and it remained dark as long as the intermission lasted, etc.[2]

These last observations later became the starting point for new ideas which guided me in making investigations as to the chemical cause of the change in color of glandular blood during secretion. I shall not further describe these experiments which, moreover, I have published in detail.[3] It is enough for me to prove that scientific investigations and experimental ideas may have their birth in almost involuntary chance observations which present themselves either spontaneously or in an experiment made with a different purpose.

[2] Claude Bernard, *Leçons sur les propriétés physiologiques et les altérations pathologiques des liquides de l'organisme.* Paris, 1859. Vol. II.

[3] Claude Bernard, *Sur la quantité d'oxygène que contient le sang veineux des organes glandulaires.* (*Compt. rend. de l'Acad. des sciences.* Vol. XLVII, Sept. 6, 1858.)

Let me cite another case,—one in which an experimenter produces an observation and voluntarily brings it to birth. This case is,
so to speak, included in the preceding case; but it differs from it in
this, that, instead of waiting for an observation to present itself by
chance in fortuitous circumstances, we produce it by experiment.
Returning to Bacon's comparison, we might say that an experimenter,
in this instance, is like a hunter who, instead of waiting quietly for
game, tries to make it rise, by beating up the locality where he
assumes it is. We use this method whenever we have no preconceived idea in respect to a subject as to which previous observations
are lacking. So we experiment to bring to birth observations which
in turn may bring to birth ideas. This continually occurs in medicine when we wish to investigate the action of a poison or of some
medicinal substance, on an animal's economy; we make experiments
to see, and we then take our direction from what we have seen.

Fourth example.—In 1845, Monsieur Pelouze gave me a toxic
substance, called *curare,* which had been brought to him from America. We then knew nothing about the physiological action of this
substance. From old observations and from the interesting accounts
of Alex. von Humboldt and of Roulin and Boussingault, we knew only
that the preparation of this substance was complex and difficult, and
that it very speedily kills an animal if introduced under the skin.
But from the earlier observations, I could get no idea of the mechanism of death by curare; to get such an idea I had to make fresh observations as to the organic disturbances to which this poison might lead.
I therefore made experiments *to see* things about which I had absolutely no preconceived idea. First, I put curare under the skin of
a frog: it died after a few minutes; I opened it at once, and in this
physiological autopsy I studied in succession what had become of
the known physiological properties of its various tissues. I say
physiological autopsy purposely, because no others are really instructive. The disappearance of physiological properties is what explains death, and not anatomical changes. Indeed, in the present
state of science, we see physiological properties disappear in any
number of cases without being able to show, by our present means
of observation, any corresponding anatomical change; such, for
example, is the case with curare. Meantime, we shall find examples,
on the contrary, in which physiological properties persist, in spite
of very marked anatomical changes with which the functions are by

no means incompatible. Now in my frog poisoned with curare, the heart maintained its movements, the blood was apparently no more changed in physiological properties than the muscles, which kept their normal contractility. But while the nervous system had kept its normal anatomical appearance, the properties of the nerves had nevertheless completely disappeared. There were no movements, either voluntary or reflex, and when the motor nerves were stimulated directly, they no longer caused any contraction in the muscles. To learn whether there was anything accidental or mistaken in this first observation, I repeated it several times and verified it in various ways; for when we wish to reason experimentally, the first thing necessary is to be a good observer and to make quite certain that the starting point of our reasoning is not a mistake in observation. In mammals and in birds, I found the same phenomena as in frogs, and disappearance of the physiological properties of the motor nervous system became my constant fact. Starting from this well established fact, I could then carry analysis of the phenomena further and determine the mechanism of death from curare. I still proceeded by reasonings analogous to those quoted in the above example, and, from idea to idea and experiment to experiment, I progressed to more and more definite facts. I finally reached this general proposition, that *curare causes death by destroying all the motor nerves, without affecting the sensory nerves.*[4]

In cases where we make an experiment in which both preconceived idea and reasoning seem completely lacking, we yet necessarily reason by syllogism without knowing it. In the case of curare, I instinctively reasoned in the following way: no phenomenon is without a cause, and consequently no poisoning without a physiological lesion peculiar or proper to the poison used; now, thought I, curare must cause death by an activity special to itself and by acting on certain definite organic parts. So by poisoning an animal with curare and by examining the properties of its various tissues immediately after death, I can perhaps find and study the lesions peculiar to it.

The mind, then, is still active here, and an experiment in order to see is included, nevertheless, in our general definition of an experiment (p. 10). In every enterprise, in fact, the mind is always rea-

[4] Cf. Claude Bernard, *Leçons sur les effets des substances toxiques;* Paris, 1857; *Du curare.* (*Revue des Deux Mondes*, Sept. 1, 1864.)

soning, and, even when we seem to act without a motive, an instinctive logic still directs the mind. Only we are not aware of it, because we begin by reasoning before we know or say that we are reasoning, just as we begin by speaking before we observe that we are speaking, and just as we begin by seeing and hearing before we know what we see or what we hear.

Fifth example.—About 1846, I wished to make experiments on the cause of poisoning with carbon monoxide. I knew that this gas had been described as toxic, but I knew literally nothing about the mechanism of its poisoning; I therefore could not have a preconceived opinion. What, then, was to be done? I must bring to birth an idea by making a fact appear, i.e., make another experiment to see. In fact I poisoned a dog by making him breathe carbon monoxide and after death I at once opened his body. I looked at the state of the organs and fluids. What caught my attention at once was that its blood was scarlet in all the vessels, in the veins as well as the arteries, in the right heart as well as in the left. I repeated the experiment on rabbits, birds and frogs, and everywhere I found the same scarlet coloring of the blood. But I was diverted from continuing this investigation, and I kept this observation a long time unused except for quoting it in my course à propos of the coloring of blood.

In 1856, no one had carried the experimental question further, and in my course at the Collège de France on toxic and medicinal substances, I again took up the study of poisoning by carbon monoxide which I had begun in 1846. I found myself then in a confused situation, for at this time I already knew that poisoning with carbon monoxide makes the blood scarlet in the whole circulatory system. I had to make hypotheses, and establish a preconceived idea about my first observation, so as to go ahead. Now, reflecting on the fact of scarlet blood, I tried to interpret it by my earlier knowledge as to the cause of the color of blood. Whereupon all the following reflections presented themselves to my mind. The scarlet color, said I, is peculiar to arterial blood and connected with the presence of a large proportion of oxygen, while dark coloring belongs with absence of oxygen and presence of a larger proportion of carbonic acid; so the idea occurred to me that carbon monoxide, by keeping venous blood scarlet, might perhaps have prevented the oxygen from changing into carbonic acid in the capillaries. Yet it seemed hard to understand how that could be the cause of death. But still keeping on with my inner preconceived reasoning, I added: If that is true, blood taken from the veins of animals poisoned with carbon monoxide should be like arterial blood in containing oxygen; we must see if that is the fact.

Following this reasoning, based on interpretation of my observation, I tried an experiment to verify my hypothesis as to the persistence of oxygen in the venous blood. I passed a current of hydrogen through scarlet venous blood taken from an animal poisoned with carbon monoxide, but I could not liberate the oxygen as usual. I tried to do the same with arterial blood; I had no greater success. My preconceived idea was therefore false. But the impossibility of getting oxygen from the blood of a dog poisoned with carbon monoxide was a second observation which suggested a fresh hypothesis. What could have become of the oxygen in the blood? It had not changed into carbonic acid, because I had not set free large quantities of that gas in passing a current of hydrogen through the blood of the poisoned animals. Moreover, that hypothesis was contrary to the color of the blood. I exhausted myself in conjectures about how carbon monoxide could cause the oxygen to disappear from the blood; and as gases displace one another I naturally thought that the carbon monoxide might have displaced the oxygen and driven it out of the blood. To learn this, I decided to vary my experimentation by putting the blood in artificial conditions that would allow me to recover the displaced oxygen. So I studied the action of carbon monoxide on blood experimentally. For this purpose I took a certain amount of arterial blood from a healthy animal; I put this blood on the mercury in an inverted test tube containing carbon monoxide; I then shook the whole thing so as to poison the blood sheltered from contact with the outer air. Then, after an interval, I examined whether the air in the test-tube in contact with the poisoned blood had been changed, and I noted that the air thus in contact with the blood had been remarkably enriched with oxygen, while the proportion of carbon monoxide was lessened. Repeated in the same conditions, these experiments taught me that what had occurred was an exchange, volume by volume, between the carbon monoxide and the oxygen of the blood. But the carbon monoxide, in displacing the oxygen that it had expelled from the blood, remained chemically combined in the blood and could no longer be displaced either by oxygen or by other gases. So that death came through death of the molecules of blood, or in other words by stopping their exercise of a physiological property essential to life.

This last example, which I have very briefly described, is complete; it shows from one end to the other, how we proceed with the experimental method and succeeded in learning the immediate cause of phenomena. To begin with I knew literally nothing about the mechanism of the phenomenon of poisoning with carbon monoxide. I undertook an experiment to see, i.e., to observe. I made a preliminary observation of a special change in the coloring of blood. I interpreted this observation, and I made an hypothesis which proved false. But the experiment provided me with a second observation about which I reasoned anew, using it as a starting point for making a new hypothesis as to the mechanism, by which the oxygen in the blood was removed. By building up hypotheses, one by one, about the facts as I observed them, I finally succeeded in showing that carbon monoxide replaces oxygen in a molecule of blood, by combining with the substance of the molecule. Experimental analysis, here, has reached its goal. This is one of the cases, rare in physiology, which I am happy to be able to quote. Here the immediate cause of the phenomenon of poisoning is found and is translated into a theory which accounts for all the facts and at the same time includes all the observations and experiments. Formulated as follows, the theory posits the main facts from which all the rest are deduced: Carbon monoxide combines more intimately than oxygen with the hemoglobin in a molecule of blood. It has quite recently been proved that carbon monoxide forms a definite combination with hemoglobin.[5] So that the molecule of blood, as if petrified by the stability of the combination, loses its vital properties. Hence everything is logically deduced: because of its property of more intimate combination, carbon monoxide drives out of the blood the oxygen essential to life; the molecules of blood become inert, and the animal dies, with symptoms of hemorrhage, from true paralysis of the molecules.

But when a theory is sound and indeed shows the real and definite physico-chemical cause of phenomena, it not only includes the observed facts but predicts others and leads to rational applications that are logical consequences of the theory. Here again we

[5] Hoppe-Seyler, *Handbuch der physiologisch- und pathologisch-chemischen Analyse.* Berlin, 1865.

meet this criterion. In fact, if carbon monoxide has the property of driving out oxygen by taking its place in combining with a molecule of blood, we should be able to use the gas to analyze the gases in blood, and especially for determining oxygen. From my experiments I deduced this application which has been generally adopted to-day.[6] Applications of this property of carbon monoxide have been made in legal medicine for finding the coloring matter of blood; and from the physiological facts described above we may also already deduce results connected with hygiene, experimental pathology, and notably with the mechanism of certain forms of anemia.

As in every other case, all the deductions from the theory doubtless still require experimental verification; and logic does not suffice. But this is because the conditions in which carbon monoxide acts on the blood may present other complex circumstances and any number of details which the theory cannot yet predict. Otherwise, as we have often said (p. 29), we could reach conclusions by logic alone, without any need of experimental verifications. Because of possible unforeseen and variable new elements in the conditions of a phenomenon, logic alone can in experimental science never suffice. Even when we have a theory that seems sound, it is never more than relatively sound, and it always includes a certain proportion of the unknown.

II. When the Starting Point of Experimental Research Is an Hypothesis or a Theory

We have already said (p. 25) and we shall see further on, that in noting an observation we must never go beyond facts. But in making an experiment, it is different. I wish to show that hypotheses are indispensable, and that they are useful, therefore, precisely because they lead us outside of facts and carry science forward. The object of hypotheses is not only to make us try new experiments; they also often make us discover new facts which we should not have perceived without them. In the preceding examples, we saw that we can start from a particular fact and rise one by one to more general ideas, i.e., to a theory. But as we have just seen, we can also sometimes

---

[6] Claude Bernard, *De l'emploi de l'oxyde de carbone pour la détermination de l'oxygène du sang* (*Compt. rend. de l'Acad. des sciences*, Meeting of Sept. 6, 1858, Vol. XLVII).

start with an hypothesis deduced from a theory. Though we are dealing in this case with reasoning logically deduced from a theory, we have an hypothesis that must still be verified by experiment. Indeed, theories are only an assembling of the earlier facts, on which our hypothesis rests, and cannot be used to demonstrate it experimentally. We said that, in this instance, we must not submit to the yoke of theories, and that keeping our mental independence is the best way to discover the truth. This is proved by the following examples.

First example.—In 1843, in one of my first pieces of work, 1 undertook to study what becomes of different alimentary substances in nutrition. As I said before, I began with sugar, a definite substance that is easier than any other to recognize and follow in the bodily economy. With this in view, I injected solutions of cane sugar into the blood of animals, and I noted that even when injected in weak doses the sugar passed into the urine. I recognized later that, by changing or transforming sugar, the gastric juice made it capable of assimilation, i.e., of destruction in the blood.[7]

Thereupon I wished to learn in what organ the nutritive sugar disappeared, and I conceived the hypothesis that sugar introduced into the blood through nutrition might be destroyed in the lungs or in the general capillaries. The theory, indeed, which then prevailed and which was naturally my proper starting point, assumed that the sugar present in animals came exclusively from foods, and that it was destroyed in animal organisms by the phenomena of combustion, i.e., of respiration. Thus sugar had gained the name of *respiratory nutriment*. But I was immediately led to see that the theory about the origin of sugar in animals, which served me as a starting point, was false. As a result of the experiments which I shall describe further on, I was not indeed led to find an organ for destroying sugar, but, on the contrary, I discovered an organ for making it, and I found that all animal blood contains sugar even when they do not eat it. So I noted a new fact, unforeseen in theory, which men had not noticed, doubtless because they were under the influence of contrary theories which they had too confidently accepted. I therefore abandoned my hypothesis on the spot, so as to pursue the unexpected result which has since become the fertile

[7] Claude Bernard, Thesis for doctorate in medicine, Paris, 1843.

origin of a new path for investigation and a mine of discoveries that is not yet exhausted.

In these researches I followed the principles of the experimental method that we have established, i.e., that, in presence of a well-noted, new fact which contradicts a theory, instead of keeping the theory and abandoning the fact, I should keep and study the fact, and I hastened to give up the theory, thus conforming to the precept which we proposed in the second chapter: "When we meet a fact which contradicts a prevailing theory, we must accept the fact and abandon the theory, even when the theory is supported by great names and generally accepted."

We must therefore distinguish, as we said, between principles and theories, and never believe absolutely in the latter. We had a theory here which assumed that the vegetable kingdom alone had the power of creating the individual compounds which the animal kingdom is supposed to destroy. According to this theory, established and supported by the most illustrious chemists of our day, animals were incapable of producing sugar in their organisms. If I had believed in this theory absolutely, I should have had to conclude that my experiment was vitiated by some inaccuracy; and less wary experimenters than I might have condemned it at once, and might not have tarried longer at an observation which could be theoretically suspected of including sources of error, since it showed sugar in the blood of animals on a diet that lacked starchy or sugary materials. But instead of being concerned about the theory, I concerned myself only with the fact whose reality I was trying to establish. By new experiments and by means of suitable counterproofs, I was thus led to confirm my first observation and to find that the liver is the organ in which animal sugar is formed in certain given circumstances, to spread later into the whole blood supply and into the tissues and fluids.

Animal glycogenesis which I thus discovered, i.e., the power of producing sugar, possessed by animals as well as vegetables, is now an acquired fact for science; but we have not yet fixed on a plausible theory accounting for the phenomenon. The fresh facts which I made known are the source of numerous studies and many varied theories in apparent contradiction with each other and with my own. When entering on new ground we must not be afraid to express even risky ideas so as to stimulate research in all directions. As Priestley put it, we must not remain inactive through false modesty based on fear of being mistaken. So I made more or less hypothetical theories of glycogenesis; after mine came others; my theories, like other men's, will live the allotted life of necessarily very partial and temporary theories at the opening of a new series of investigations; they will be replaced later by others, embodying a more advanced stage of the question, and so on. Theories are like a stairway; by climbing, science widens its horizon more and more, because theories embody and necessarily include proportionately more facts as they advance. Progress is achieved by exchanging our theories for new ones which go further than the old, until we find one based on a larger number of facts. In the case which now concerns us, the question is not one of condemning the old to the advantage of a more recent theory. What is important is having opened a new road; for well-observed facts, though brought to light by passing theories, will never die; they are the material on which alone the house of science will at last be built, when it has facts enough and has gone sufficiently deep into the analysis of phenomena to know their law or their causation.

To sum up, theories are only hypotheses, verified by more or less numerous facts. Those verified by the most facts are the best; but even then they are never final, never to be absolutely believed. We have seen in the preceding examples that if we had had complete confidence in the prevailing theory of the destruction of sugar in animals, and if we had only had its confirmation in view, we should probably not have found the road to the new facts which we met. It is true that an hypothesis based on a theory produced the experiment; but as soon as the results of the experiment appeared, theory and hypothesis had to disappear, for the experimental facts were now just an observation, to be made without any preconceived idea (p. 21).

In sciences as complex and as little developed as physiology, the great principle is therefore to give little heed to hypotheses or theories and always to keep an eye alert to observe everything that appears in every aspect of an experiment. An apparently accidental and inexplicable circumstance may occasion the discovery of an important new fact, as we shall see in the continuation of the example just noted.

Second example (Sequel to the Last).—After finding, as I said above, that there is sugar in the livers of animals in their normal state, and with every sort of nutriment, I wished to learn the proportion of this substance and its variation in certain physiological and pathological states. So I began to estimate the sugar in the livers of animals placed in various physiologically defined circumstances. I always made two determinations of carbohydrate for the same liver tissue. But pressed for time one day, it happened that I could not make my two analyses at the same moment; I quickly made one determination just after the animal's death and postponed the other analysis till next day. But then I found much larger amounts of sugar than those which I got the night before with the same material. I noticed, on the other hand, that the proportion of sugar, which I had found just after the animal's death the night before, was much smaller than I had found in the experiments which I had announced as giving the normal proportion of liver sugar. I did not know how to account for this singular variation, got with the same liver and the same method of analysis. What was to be done? Should I consider two such discordant determinations as an unsuccessful experiment and take no account of them? Should I take the mean between these experiments? More than one experimenter might have chosen this expedient to get out of an awkward situation. But I disapprove of this kind of action for reasons which I have given elsewhere. I said, indeed, that we must never neglect anything in our observation of fact, and I consider it indispensable, never to admit the existence of an unproved source of error in an experiment and always to try to find a reason for the abnormal circumstances that we observe. Nothing is accidental, and what seems to us accident is only an unknown fact whose explanation may furnish the occasion for a more or less important discovery. So it proved in this case.

I wished, in fact, to learn the reason for my having found two such different values in the analysis of my rabbit's liver. After assuring myself that there was no mistake connected with the method of analysis, after noting that all parts of the liver were practically equally rich in sugar, there remained to be studied only the elapsed time between the animal's death and the time of my second determination. Without ascribing much importance to it, up to that time I had made my experiments a few hours after the animal's death; now for the first time I was in the situation of making one determination only a few minutes after death and postponing the other till next day, i.e., twenty-four hours later. In physiology, questions of time are always very important because organic matter passes through numerous and incessant changes. Some chemical change might therefore have taken place in the liver tissue. To make sure, I made a series of new experiments which dispelled every obscurity by showing me that liver tissue becomes more and more rich in sugar for some time after death. Thus we may have a very variable amount of sugar according to the moment when we make our examination. I was therefore led to correct my old determination and to discover the new fact that considerable amounts of sugar are produced in animals' livers after death. For instance, by forcibly injecting a current of cold water through the hepatic vessels and passing it through a liver that was still warm, just after an animal's death, I showed that the tissue was completely freed from the sugar which it contained; but next day or a few hours later, if we keep the washed liver at a mild temperature, we again find its tissue charged with a large amount of sugar produced after it was washed.[8]

Once in possession of the first discovery that sugar is formed in animals after death as during life, I wished to carry my study of this singular phenomenon further; I was then led to find that sugar is produced in the liver with the help of an enzyme reacting on an amylaceous substance which I isolated and which I called *glycogenous matter,* so that I succeeded in proving in the most clear-cut way that sugar is formed in animals by a mechanism in every respect like the mechanism found in vegetables.

This second series of facts embodied results, which are also firmly acquired for science, and which have greatly advanced our knowledge of glycogenesis in animals. I have just very briefly told how these facts were discovered, and how they started with an experimental circumstance that was apparently inconsequential. I quote this case so as to prove that we must never neglect anything in experimental research, for every accident has a necessary cause. We must, therefore, never be too much absorbed by the thought we are pursuing, nor deceive ourselves about the value of our ideas or scientific theories; we must always keep our eyes open for every event, the mind doubting and independent (p. 80), ready to study whatever presents itself and to let nothing go without seeking its reason. In a word,

---

[8] Claude Bernard, *Sur le mécanisme de la formation du sucre dans le foie* (*Comptes rendus par l'Acad. des sciences*, Sept. 24, 1855, and *Comptes rendus de l'Acad. des sciences*, March 23, 1857).

we must be in an intellectual attitude which seems paradoxical but which, in my opinion, expresses the true spirit of an investigator. We must have robust faith and not believe. Let me explain myself by saying that in science we must firmly believe in principles, but must question formulæ; on the one hand, indeed, we are sure that determinism exists, but we are never certain we have attained it. We must be immovable as to the principles of experimental science (determinism), but must not absolutely believe in theories. The aphorism which I just uttered is sustained by what we expounded elsewhere (p. 67), to wit, that for experimental science principles are in our mind, while formulæ are external things. In practical matters, we are indeed forced to tolerate the belief that truth (at least temporary truth) is embodied in a theory or a formula. But in scientific experimental philosophy those who put their faith in formulæ and theories are wrong. All human science consists in seeking the true formula and true theory. We are always approaching it; but shall we ever find it completely? This is not the place to go into an explanation of philosophic ideas: let us return to our subject and pass on to a fresh experimental example.

Third example.—About the year 1852, my studies led me to make experiments on the influence of the nervous system on the phenomena of nutrition and temperature regulation. It had been observed in many cases that complex paralyses with their seat in the mixed nerves are followed, now by a rise and again by a fall of temperature in the paralyzed parts. Now this is how I reasoned, in order to explain this fact, basing myself first on known observations and then on prevailing theories of the phenomena of nutrition and temperature regulation. Paralysis of the nerves, said I, should lead to cooling of the parts by slowing down the phenomena of combustion in the blood, since these phenomena are considered as the cause of animal heat. On the other hand, anatomists long ago noticed that the sympathetic nerves especially follow the arteries. So, thought I inductively, in a lesion of a mixed trunk of nerves, it must be the sympathetic nerves that produce the slowing down of chemical phenomena in capillary vessels, and their paralysis that then leads to cooling the parts. If my hypothesis is true, I went on, it can be verified by severing only the sympathetic, vascular nerves leading to a special part, and sparing the others. I should then find the part cooled by paralysis of the vascular nerves, without loss of either motion or sensation, since the ordinary motor and sensory nerves would still be intact. To carry out my experiment, I therefore sought a suitable experimental method that would allow me to sever only the vascular nerves and to spare the others. Here the choice of animals was important in solving the problem (p. 122); for in certain animals, such as rabbits and horses, I found that the anatomical arrangement isolating the cervical sympathetic nerve made this solution possible.

Accordingly, I severed the cervical sympathetic nerve in the neck of a rabbit, to control my hypothesis and see what would happen in the way of change of temperature on the side of the head where this nerve branches out. On the basis of a prevailing theory and of earlier observation, I had been led, as we have just seen, to make the hypothesis that the temperature should be reduced. Now what happened was exactly the reverse. After severing the cervical sympathetic nerve about the middle of the neck, I immediately saw in the whole of the corresponding side of the rabbit's head a striking hyperactivity in the circulation, accompanied by increase of warmth. The result was therefore precisely the reverse of what my hypothesis, deduced from theory, had led me to expect; thereupon I did as I always do, that is to say, I at once abandoned theories and hypothesis, to observe and study the fact itself, so as to define the experimental conditions as precisely as possible. To-day my experiments on the vascular and thermo-regulatory nerves have opened a new path for investigation and are the subject of numerous studies which, I hope, may some day yield really important results in physiology and pathology.[9] This example, like the preceding ones, proves that in experiments we may meet with results different from what theories and hypothesis lead us to expect. But I wish to call more special attention to this third example, because it gives us an important lesson, to wit: without the original guiding hypothesis, the experimental fact which contradicted it would never have been perceived. Indeed, I was not the first experimenter to cut this part of the cervical sympathetic nerve in living animals. Pourfour du Petit performed the experiment at the beginning of the last century and discovered the

---

[9] Claude Bernard, *Recherches expérimentales sur le grand sympathique, etc.* (*Mémoires de la Société de biologie*, Vol. V, 1853). *Sur les nerfs vasculaires et caloriques du grand sympathique* (*Comptes rendus de l'Acad. des sciences*, 1852, Vol. XXXIV; 1862, Vol. LV).

nerve's action on the pupil, by starting from an anatomical hypothesis according to which this nerve was supposed to carry animal spirits to the eye.[10] Many physiologists have since repeated the same operation, with the purpose of verifying or explaining the changes in the eye which Pourfour du Petit first described. But none of them noticed the local temperature phenomenon, of which I speak, or connected it with the severing of the cervical sympathetic nerve, though this phenomenon must necessarily have occurred under the very eyes of all who, before me, had cut this part of the sympathetic nerve. The hypothesis, as we see, had prepared my mind for seeing things in a certain direction, given by the hypothesis itself; and this is proved by the fact that, like the other experimenters, I myself had often divided the cervical sympathetic nerve to repeat Pourfour du Petit's experiment, without perceiving the fact of heat production which I later discovered when an hypothesis led me to make investigations in this direction. Here, therefore, the influence of the hypothesis could hardly be more evident; we had the fact under our eyes and did not see it because it conveyed nothing to our mind. However, it could hardly be simpler to perceive, and since I described it, every physiologist without exception has noted and verified it with the greatest ease.

To sum up, even mistaken hypotheses and theories are of use in leading to discoveries. This remark is true in all the sciences. The alchemists founded chemistry by pursuing chimerical problems and theories which are false. In physical science, which is more advanced than biology, we might still cite men of science who make great discoveries by relying on false theories. It seems, indeed, a necessary weakness of our mind to be able to reach truth only across a multitude of errors and obstacles.

What general conclusions shall physiologists draw from the above examples? They should conclude that in the present state of biological science accepted ideas and theories embody only limited and risky truths which are destined to perish. They should consequently have very little confidence in the ultimate value of theories, but should still make use of them as intellectual tools necessary to the evolution of science and suitable for the discovery of new facts.

---

[10] Pourfour du Petit, *Mémoire dans lequel il est démontré que les nerfs intercostaux fournissent des rameaux qui portent des esprits dans les yeux* (*Histoire de l'Académie pour l'année* 1727).

The art of discovering new phenomena and of noting them accurately should to-day be the special concern of all biologists. We must establish experimental criticism by creating rigorous methods of investigation and experimentation, which will enable us to define our observations unquestionably, and thus get rid of the errors of fact which are the source of errors in theory. A man who to-day attempted a generalization for biology as a whole would prove that he had no accurate feeling for the present state of the science. To-day, the biological problem has hardly begun to be put; and, as stones must first be got together and cut, before we dream of erecting a monument, just so must the facts first be got together and prepared which are destined to create the science of living bodies. This rôle falls to experimentation; its method is fixed, but the phenomena to be analyzed are so complex that, for the moment, the true promoters of science are those who succeed in giving its methods of analysis a few principles of simplification or in introducing improvements in instruments of research. When there are enough quite clearly established facts, generalizations never keep us waiting. I am convinced that, in experimental sciences that are evolving, and especially in those as complex as biology, discovery of a new tool for observation or experiment is much more useful than any number of systematic or philosophic dissertations. Indeed, a new method or a new means of investigation increases our power and makes discoveries and researches possible which would not have been possible without its help. Thus researches as to the formation of sugar in animals could be made only after chemistry gave us reagents for recognizing sugar, which were much more sensitive than those we had before.

CHAPTER II

EXAMPLES OF EXPERIMENTAL PHYSIOLOGICAL CRITICISM

EXPERIMENTAL criticism rests on absolute principles which must guide experimenters in noting and interpreting the phenomena of nature. It will be particularly useful in the biological sciences where prevailing theories are so often propped up with false ideas or based on poorly observed facts. We shall here deal with examples recalling the principles, by virtue of which we may well judge physiological theories, and discuss the facts on which they are based. As we already know, our criterion *par excellence* is the principle of experimental determinism united with philosophic doubt. In this connection, let me again recall the fact that, in science, we must never confuse principles with theories. Principles are scientific axioms; as absolute truths, they are an immutable criterion. Theories are scientific generalizations or scientific ideas which sum up our present state of knowledge; they are always relative truths, destined to change with the progress of science. So if we posit as a basic conclusion, that we must not believe absolutely in the formulæ of science, we must, on the contrary, believe absolutely in its principles. Men who too completely believe in theories and neglect principles, take the shadow for reality; they lack any solid criterion and are liable to all the consequent sources of error. In every science, progress consists in so changing our theories as to get more and more perfect ones. Indeed, of what use would study be, if we could not change our opinions or theories? But principles and the scientific method are higher than theory; they are immutable and can never change.

Experimental criticism must therefore forearm itself, not only against belief in theories, but against being led astray by too highly valuing the words which we have created to picture to ourselves the supposed forces of nature. In every science, but in the physiological sciences more than all others, we are in danger of deceiving ourselves about words. We must never forget that our characterizations of the phenomena of nature, as mineral or vital forces, are merely figurative language by which we must not allow ourselves to be duped. The only realities are manifestations of phenomena and the conditions of these manifestations which remain to be determined; experimental criticism should never lose sight of that. In a word, experimental criticism casts doubt on everything except the principle of scientific, rational determinism in the realm of facts (pp. 52-67). It is always founded on this same base, whether we direct it against ourselves or others; that is why we shall usually present two examples in what follows, one chosen from our own researches, the other from other men's work. In science, indeed, we must not only try to criticise others, but every man of science must always be a severe critic of himself. Whenever he proffers an opinion or proposes a theory, he must be the first to try to control it by criticism and to base it on well observed and accurately determined facts.

I. The Principle of Experimental Determinism Does Not Admit of Contradictory Facts

First example.—It is now a long time since I announced an experiment which greatly surprised physiologists: the experiment consists in making an animal artificially diabetic by means of a puncture in the floor of the fourth ventricle. I was led to try this puncture as a result of theoretical considerations which I need not recall; all that we here need to know is that I succeeded at the first attempt, i.e., that I saw the first rabbit on which I operated become strikingly diabetic. But I afterward had the experience of repeating the experiment many times (eight or ten times) without getting the same result. I then found myself in presence of a positive fact and of eight or ten negative facts; yet I never thought of denying my first positive experiment in favor of the negative experiments which followed it. Thoroughly convinced that my failures were due only to not knowing the true conditions of my first experiment, I persisted in experimenting, to try to discover them. As a result, I succeeded in defining the exact place for the puncture, and showing the conditions in which the animal to be operated on should be placed; so that we can to-day reproduce artificial diabetes whenever we place ourselves in the conditions known to be necessary to its appearance.

Let me add to the above a reflection showing how many sources of error may surround physiologists in the investigation of vital phenomena. Let me assume that, instead of succeeding at once in making a rabbit diabetic, all the negative facts had first appeared; it is clear that, after failing two or three times, I should have concluded, not only that the theory guiding me was false, but that puncture of the fourth ventricle did not produce diabetes. Yet I should have been wrong. How often men must have been and still must be wrong in this way! It even seems impossible absolutely to avoid this kind of mistake. We wish to draw from this experiment another general conclusion which will be corroborated by subsequent examples, to wit, that negative facts when considered alone, never teach us anything.

Second example.—Every day we see discussions which remain profitless for science, because we are not thoroughly enough imbued with the principle that, since every fact has its own appropriate cause, a negative fact proves nothing and can never destroy a positive fact. To prove what I am setting forth, I will quote the criticisms which M. Longet formerly made on Magendie's experiments. I choose this example, on the one hand, because it is highly instructive, and, on the other, because I was involved in it, and know all the circumstances accurately. Let me begin with M. Longet's criticisms about Magendie's experiments on the properties of recurrent sensitivity in the anterior spinal roots.[1] The first objection which M. Longet makes to Magendie is that he changed his opinion as to the sensitivity of the anterior roots, holding in 1822 that the anterior roots were scarcely sensitive, and in 1839 that they were very sensitive, etc. Thereupon M. Longet exclaims: "Truth is single; from the midst of these contrasted, contradictory assertions of the same author, let the reader choose if he dare." (loc. cit., p. 22.) Finally M. Longet goes on, "M. Magendie ought at least to have told us,—to get us out of our difficulties,—which of his experiments were properly made, the 1822 experiments or those in 1839." (loc. cit., p. 23.)

These criticisms are all ill founded and completely violate the

---

[1] F. A. Longet, *Recherches cliniques et expérimentales sur les fonctions des faisceaux de la moelle épinière et des racines des nerfs rachidiens, précédées d'un Examen historique et critique des expériences faites sur ces organes depuis Sir Ch. Bell, et suivies d'autres recherches sur diverses parties du système nerveux* (*Archives générales de médecine*, 1841, 3d Series, Vol. X, p. 296 and Vol. XI p. 129).

rules of experimental scientific criticism. In fact, if Magendie in 1822 said that the anterior roots were insensitive, and if he said later in 1839 that the anterior roots are very sensitive, it was because he then found them very sensitive. We do not have to choose between the two results, as M. Longet believes; we must accept them both and merely explain and define them in their respective conditions. When M. Longet exclaims: "Truth is single," does he mean that if one of these results is true, the other must be false? By no means; they are both true, unless we say that in one case Magendie lied, and that is certainly not the critic's idea. But by virtue of the scientific principle of the determinism of phenomena, we must absolutely affirm *a priori* that in 1822 and in 1839 Magendie did not see the phenomena in identical conditions; the differences in conditions are precisely what we must seek out and define, so as to harmonize the two results and thus find the cause of variation in the phenomenon. The only objection which M. Longet might have made to Magendie was that he did not himself seek out the reason for the difference in the two results; but the criticism by exclusion that M. Longet directed against Magendie's experiments is false and, as we said, out of harmony with the principles of experimental criticism.

We cannot doubt that the above criticism is sincere and purely scientific; for in other circumstances connected with the same discussion M. Longet directed against himself the same criticism by exclusion and in his own criticism was led into the same kind of mistake as in the criticism directed against Magendie.

In 1839 M. Longet, like myself, was working in the laboratory of the Collège de France when Magendie discovered the sensitivity of the anterior spinal roots and showed that it is derived from the posterior roots and returns by the periphery, whence the name reverse sensitivity or recurrent sensitivity which he gave it. Like Magendie and me, M. Longet then saw that the anterior root was sensitive and that it was so under the influence of the posterior root, and he saw it so clearly that he claimed discovery of the latter fact for himself.[2] But later, in 1841, when Longet wished to repeat Magendie's experiment, he found no sensitiveness in the anterior root. In rather amusing circumstances, M. Longet thus found him-

---

[2] Longet, *Comptes rendus de l'Académie des sciences*, Vol. VIII, p. 787, June 3 and 10; *Comptes rendus de l'Académie des sciences*, June 4; *Gazette des hôpitaux*, June 13 and 18, 1839.

self in exactly the same position in relation to the same fact of sensitiveness in the anterior spinal roots with which he had reproached Magendie, i.e., M. Longet in 1839 saw that the anterior spinal root was sensitive and, in 1841, saw that it was insensitive. Magendie's sceptical mind was not disturbed by these seeming obscurities and contradictions; he went on experimenting and always said what he saw. M. Longet's mind, on the contrary, wished to have the truth on one side or the other; that is why he decided in favor of the 1841 experiments, i.e., the negative experiments; and here is what he said: "Though at that time (1839) I brought forward my claim to the discovery of one of these facts (recurrent sensitiveness), now that I have made many and varied experiments on this point in physiology, I combat these very facts as erroneous, whether they are regarded as Magendie's property or my own. When we have made a mistake, the service which we owe to truth requires that we should never fear retraction. I shall here only recall the insensitivity of the anterior roots and sheaves which we have so often proved, so that the reader may readily understand how meaningless are these results which, like so many others, merely encumber science and embarrass its advance."[3] After this confession, we may be sure that M. Longet is animated only by a desire to find the truth, which he proves when he says that we must never be afraid of retraction if we have made an error. I wholly share this feeling. Let me add that it is always instructive to acknowledge an error. The precept, therefore, is excellent, for we are all likely to make mistakes, except those of us who do nothing. But the first requirement in acknowledging a mistake is to prove that there is an error. It is not enough to say: I was mistaken; we must say how we were mistaken; the important point is precisely that. Now M. Longet explains nothing; he seems purely and simply to say: In 1839 I saw sensitive roots; in 1841, I saw insensitive ones more often, therefore I was mistaken in 1839. Such reasoning is inadmissible. Here, in fact, are a number of experiments in 1839, *à propos* to the sensitivity of anterior roots,—experiments in which the spinal roots were cut one by one; and to note their properties, their ends were pinched. Magendie wrote half a volume on the subject. Later when people fail to obtain the same results, the question cannot be decided simply by saying that we made a mistake the first time and are right the second time. After all, why

---

[3] Longet, loc. cit. p. 21.

should we be mistaken? Shall we say that our senses played us false
at one period and not the other? In that event we must give up
experimentation; for the first requirement of experimenters is confidence in their senses and never any doubt except as to interpretations. If we cannot find the concrete reason for an error, despite all
our efforts and all our investigations, we must suspend judgment
and meantime keep both results; but never believe that denying
positive facts can suffice even in the name of more numerous negative
facts, or *vice versa*. Negative facts, no matter how numerous they
may be, can never destroy a single positive fact. That is why pure
and simple negation is not criticism, and this method should
be absolutely rejected in science, because science is never built up
by negation.

To sum up, we must maintain the conviction that negative facts
are determined like positive facts. We posited the principle that all
experiments are successful, in that their conditions are determined;
in research into the conditions of each of these determinations lie the
lessons that teach us the law of a phenomenon; because in this way
we learn the conditions necessary to its existence and its non-exist-
ence. After witnessing Magendie's experiments in 1839 and M.
Longet's discussions in 1841, I have made this principle my guide,
when I wished to take account of the phenomena myself and to judge
the differences. I repeated the experiments, and, like Magendie and
like M. Longet, I found cases of sensitivity and cases of insensitivity
of the anterior spinal roots; but as I was convinced that the two cases
depended on different experimental conditions, I tried to define the
conditions; by dint of observation and perseverance, I finally found [4]
the conditions in which we must place ourselves to get both results.
Now that the conditions of the phenomenon are known, it is no longer
questioned. M. Longet himself [5] and all physiologists accept the fact
of recurrent sensitivity as constant in the conditions which I announced.

From what has gone before, we must therefore establish the absolute and necessary determination of phenomena as a principle of experimental criticism. This principle, when thoroughly understood,
should make us cautious about our natural tendency to contradic-

---

[4] Claude Bernard, *Leçons sur la physiologie et la pathologie du système
nerveux*, p. 32.

[5] See Longet, *Traité de physiologie*, 1860, Vol. II, p. 177.

tion. Certainly every experimenter, especially every beginner, feels a secret pleasure whenever he finds something different from what others have seen before him. His first impulse is to contradict, especially when contradicting someone high in the scientific world. We must protect ourselves against this tendency, for it is not scientific. Pure contradiction would amount to an accusation of lying, and we should avoid it because happily scientific falsifiers are rare. As such cases, moreover, have no connection with science, I need not offer any precept on the subject. I wish merely to point out here that science does not consist in proving that others are mistaken; and even if we proved that an eminent man was mistaken, that would not be a great discovery; it can be a profitable work for science only in so far as we show how he was mistaken. Indeed, great men often teach us by their errors as much as by their discoveries. I sometimes hear it said that pointing out an error is equivalent to a discovery. Yes, on condition that we bring to light a new truth by showing the source of error, in which event it is unnecessary to combat the error; it falls of itself. Thus only is criticism equivalent to a discovery; when it explains everything without denying anything and finds the correct causation of apparently contradictory facts. By such determinism everything is unified, everything becomes transparent; and as Leibnitz says, science, as it broadens, grows clear and simple.

II. The Principle of Determinism Ejects Causeless and Irrational Facts from Science

We said elsewhere (p. 54) that our reason scientifically includes the determinate and the indeterminate but that it cannot accept the indeterminable, because that would be nothing but accepting the marvellous, occult or supernatural which should be absolutely banished from all experimental science. The result is that a fact gains scientific value only through knowledge of its causation. A crude fact is not scientific, and a fact whose causation is irrational should also be ejected from science. Indeed, if an experimenter must submit his ideas to the criterion of facts, I do not acknowledge that he must submit his reason; for then he would extinguish the torch of his inner criterion and would necessarily fall into the realm of the indeterminable, i.e., the occult and the marvellous. In science, many crude facts are doubtless still incomprehensible; I do not mean to conclude that we should willfully reject all these facts, I wish simply to say that they should be held in reserve, for a time, as crude facts, and not introduced into science, i.e., into experimental reasoning, until their necessary conditions are defined in terms of rational determinism. Otherwise, our experimental reasoning would be continually halted or else inevitably led into the absurd.

The following examples, among many others, will prove what I assert.

First example.—A few years ago,[6] I experimented on the influence of ether on intestinal secretions. Now à propos of this, I happened to observe that injecting ether into the intestinal canal of a dog kept without food even for several days gave rise to splendid white lymphatics, absolutely like those in an animal actively digesting mixed food in which there is fat. After frequent repetitions, the fact was unquestionable. But what meaning could be ascribed to it? What reasoning was possible about its cause? Should we say: ether causes secretion of chyme? This is a fact. But that would be absurd, since there was no food in the intestine. As we see, reason rejected this causation as irrational and absurd in our present state of knowledge. I therefore tried to find the reason for this incomprehensible fact, and I finally saw that there was a source of error, because the ether dissolved the oil lubricating the piston of the syringe with which it was injected into the stomach; when the ether was injected with a glass pipette, instead of a syringe, it no longer produced the phenomenon. The irrationality of the fact, therefore, led me to see a priori that it must be false, and that it could not be used as a basis for scientific reasoning. Otherwise I should not have found the curious source of error located in the piston of a syringe. But when the source of error was once recognized, everything was explained, and the fact became rational in this sense, that the chyle was produced there, as everywhere else, by the absorption of fat; only the ether stimulated absorption and made the phenomenon more evident.

Second example.—Able and accurate experimenters[7] had seen

---

[6] Claude Bernard, Leçons sur les effets des substances toxiques et médicamenteuses, p. 428.

[7] Vulpian, Comptes rendus et Mémoires de la Société de biologie, 1854, p. 133; 1856, p. 125; 1858, 2d Series, Vol. V, Paris, 1859, p. 113; 1864.

that the venom of a toad speedily poisons frogs and other animals, while it has no effect on the toad himself. Here, in fact, is a quite simple experiment that seems to prove it: if we take venom on a lancet from the parotid glands of a domestic toad and insert it under the skin of a frog or a bird, the animals soon perish; while if we insert the same amount of venom under the skin of a toad of about the same size, he does not die of it, indeed he feels no effect. Here again is a crude fact which could become scientific only on condition that we learn how venom acts on a frog, and why it does not act on a toad. To do this, I necessarily studied the mechanism of death, for special circumstances might be encountered which would explain the difference in results on the frog and on the toad. Thus a special arrangement of the nostrils and the epiglottis explains very well why, for example, section of the two facial nerves is mortal in horses and not in other animals. But this exceptional fact is rational; it confirms the rule, as we say, in that it makes no change fundamentally in the nervous paralysis which is the same in all animals. There was nothing of the kind in the case with which we are concerned: study of the mechanism of death by toad's venom led to the conclusion that toad's venom kills by stopping the heart in frogs, while it does not act on a toad's heart. Now, in logic, we should necessarily have to admit that the muscular fibres of a toad's heart have a different nature from those of a frog's heart, since the poison which acts on the former does not act on the latter. That was impossible: for admitting that organic units identical in structure and in physiological characteristics are no longer identical in the presence of a toxic action identically the same would prove that phenomena have no necessary causation; and thus science would be denied. Pursuant to these ideas, I rejected the above-mentioned fact as irrational, and decided to repeat the experiments, even though I did not doubt their accuracy as crude fact. I then saw [8] that toad's venom easily kills frogs with a dose that is wholly insufficient for a toad, but that the latter is nevertheless poisoned if we increase the dose enough. So that the difference described was reduced to a question of quantity and did not have the contradictory meaning that might be ascribed to it. The irrationality of the fact was, therefore, again what led me to ascribe to it another meaning.

[8] Claude Bernard, *Cours de pathologie expérimentale* (*Medical Times*, 1860).

III. The Principle of Determinism Requires Comparative Determination of Facts

We have just seen that our reason forces us to reject apparently causeless facts and leads us to criticise them so as to find for them a rational meaning before using them in experimental reasoning. But since criticism, as we said, rests at once on reason and on philosophic doubt, it follows that a simple and logical appearance is not enough to make us accept an experimental fact; we should still doubt and by a counter experiment should see whether the rational appearance is not misleading. This is an absolutely strict precept, especially in medical science which by its complexity conceals additional sources of error. I have elsewhere (p. 55) described the experimental character of counterproofs; I will not return to that subject; I wish merely to point out here that, even when a fact seems logical, i.e., rational, we are never justified in omitting a counterproof or counter experiment, so that I consider this precept a kind of *order* which we must blindly follow even in cases which seem the clearest and most rational. I am going to quote two examples which will show the necessity of thus making a comparative experiment always and in spite of everything.

First example.—I explained before (p. 164) how I was once led to study the part played by sugar in nutrition and to investigate the mechanism by which this nutritive principle was destroyed in the organism. To solve this problem, I had to hunt for sugar in the blood and follow it into the intestinal vessels which absorbed it, until I could note the place where it disappeared. To carry out my experiment, I gave a dog sweetened milk soup; then I sacrificed the animal during digestion and found that the blood in the superhepatic vessels, which hold all the blood of the intestinal organs and the liver, contained sugar. It was quite natural and, as we say, logical to think that the sugar found in the superhepatic vessels was the same that I had given the animal in his soup. I am certain indeed that more than one experimenter would have stopped at that and would have considered it superfluous, if not ridiculous, to make a comparative experiment. However, I made a comparative experiment, because I was convinced of its absolute necessity on principle: which means that I am convinced that we must always doubt in physiology, even in cases where doubt seems least allowable. However, I must add that a comparative experiment was also required here by another circumstance, viz., that I used the reduction of copper as a test for sugar. This, however, is an empirical characteristic of sugar which might be shown by substances still unknown in the bodily economy. But, even apart from that, I repeat, a comparative experiment would have had to be made as an experimental necessity; for this very case proves that we can never foresee its importance.

So for comparison with the dog fed on sugary soup, I took another dog to which I gave meat to eat, being careful moreover to exclude all sugary or starchy material from its diet; then I sacrificed the animal during digestion and examined comparatively the blood in its superhepatic veins. Great was my astonishment at finding that the blood of the animal which had not eaten any also contained sugar.

We therefore see that comparative experiment led me to the discovery that sugar is constantly present in the blood of the superhepatic veins, no matter what the animal's diet may be. You may imagine that I then abandoned all hypotheses about destruction of sugar, to follow this new and unexpected fact. I first excluded all doubt of its existence by repeated experiments, and I noted that sugar also existed in the blood of fasting animals. But if benefits are linked with comparative experiment, not performing them has its necessary annoyances. This is proved by the following example:

Second example.—Magendie once made investigations on the uses of the cerebrospinal fluid and was led to the conclusion that removing this fluid produces a kind of titubation in animals and a characteristic disturbance in their motions. Indeed, after uncovering the occipito-atloidian membrane, if we pierce it to let the cerebrospinal fluid run out, we notice that the animal is seized with peculiar motor disturbances. Apparently nothing could be simpler or more natural than the influence on their motions of removal of the cerebrospinal fluid; yet this was an error, and Magendie told me how another experimenter chanced to find it. After cutting the neck muscles, this experimenter was interrupted in his experiment at the moment when he had just laid bare the occipito-atloidian membrane. Now when he came back, to go on with his experiment, he saw that the simple preliminary operation had produced the same titubation, though the cephalorachidian fluid had not been removed. What was merely the result of severing the neck muscles had therefore been attributed to removal of the cerebrospinal fluid. Comparative experiment would obviously have solved the difficulty. In this case, two animals, as we have said, ought to be placed in the same conditions save one, that is, the occipito-atloidian membrane should be laid bare in both animals, and it should be pierced, to let the fluid flow out, in only one of them; then it would be possible to judge by comparison and thus ascertain the precise part which the removal of the fluid plays in the disturbances. I might quote a great many errors into which able experimenters have fallen by neglecting the precept about comparative experiment. Only, as the examples that I have quoted prove, it is often hard to know in advance whether comparative experiment is necessary or not; and so I repeat that, to avoid all annoyance, we should accept comparative experiment as a veritable command, to be executed even when useless, so as not to be missing when it is necessary. Comparative experiments are sometimes made, now on two animals, or for greater accuracy, on two similar organs in the same animal. Thus, at the time when I wished to judge the influence of certain substances on the glycogen of the liver, I could never find two animals comparable in this respect, even by putting them in exactly similar dietary circumstances, i.e., without food for the same number of days. According to their age, sex, plumpness, etc., animals bear starvation better or worse and destroy more glycogen or less, so that I could never be sure that the differences I found were the result of difference in diet. To remove this source of error, I was forced to make the whole experiment on the same animal, by taking away a preliminary piece of its liver before the dietary injection, and another afterward. So when we want to see the influence of contraction on the metabolism of the muscle in a frog, we have to compare both members of a single animal, because in this respect two frogs are not always comparable.

IV. Experimental Criticism Should Bear on Facts Alone and Never on Words

At the beginning of this chapter, I said that we are often deceived by false values ascribed to words. I wish to explain my idea by examples.

First example.—In 1859 I made a report to the Philomathic Society, in which I discussed Brodie's and Magendie's experiments on ligature of the bile duct, and I showed that the divergent results which the two experimenters reached, depended on the fact that one operated on dogs and tied only the bile duct, while the other operated on cats and, without suspecting it, included in his ligature, both the bile duct and a pancreatic duct. Thus I showed the reason for the difference in the results they reached, and I concluded that, in physiology as everywhere else, experiments are rigorous and give identical results whenever we operate in exactly similar conditions.

À propos of this, a member of the society took the floor to attack my conclusions; it was Gerdy, surgeon at the Charité, professor in the faculty of medicine, and known through various works on surgery and physiology. "Your anatomical explanation of Brodie's and Magendie's experiments," said he, "is correct, but I cannot accept your general conclusion. You say, in fact, that the results of experiments in physiology are identical; I deny it. Your conclusion would be correct for inert nature, but it cannot be true for living nature. Whenever life enters into phenomena," he went on, "conditions may be as similar as we please; the results may still be different." To support his opinion, Gerdy cited cases of individuals with the same disease, to whom he had given the same drugs, with different results. He also recalled cases of like operations for the same disease, but followed by cure in one case and death in another. These differences, according to him, all depended on life itself altering the results, though the experimental conditions were the same; but this could not happen, he thought, in phenomena of inert bodies, into which life does not enter. Opposition to these ideas was prompt and general in the Philomathic Society. Everyone pointed out to Gerdy that his opinions were nothing less than a denial of biological science; and that, in the cases of which he spoke, he completely deceived himself as to the identity of conditions, in this sense, that the diseases which he regarded as similar or identical were not in the least alike, and that he attributed to the influence of life what should be accounted for by our ignorance about phenomena as complex as those of pathology. Gerdy continued to maintain that life had the effect of altering phenomena so as to make them differ in different individuals, even when the conditions in which they took place were identical. Gerdy believed that one man's vitality was not the same as another's, and that there must therefore be between individuals differences impossible to define. He would not give up his ideas; he entrenched himself behind the word vitality and could not be made to understand that it was only a word, devoid of meaning and corresponding to nothing; and that saying that something was due to vitality amounted to calling it unknown.

In fact, we are often duped by such words as life, death, health, disease, idiosyncrasy. We think we have explained when we say that a phenomenon is due to a vital influence, a morbid influence, or an individual idiosyncrasy. We must really learn, however, that vital phenomenon means only a phenomenon peculiar to living beings, whose cause we do not yet know; for I think that every phenomenon, called vital to-day, must sooner or later be reduced to definite properties of organized or organic matter. We may, of course, use the expression vitality as chemists use the word affinity, but knowing that fundamentally there are only phenomena and conditions of phenomena which we must learn; when the conditions necessary to phenomena are known, then occult, vital and mineral forces will disappear.

I am very happy to be in perfect harmony on this point with my colleague and friend, M. Henri Sainte-Claire Deville. This will be seen in the following words spoken by M. Sainte-Claire Deville in explaining his splendid discoveries on the effects of high temperatures to the Chemical Society of Paris.[9] "We must not conceal from ourselves the fact that studying primary causes of the phenomena which we observe and measure has its grave dangers. Escaping exact definition, independent of particular facts, it leads us, much oftener than we think, really to beg the question, and to content ourselves with specious explanations that cannot withstand severe criticism. Chiefly affinity, defined as the force which presides over chemical combinations, has long been and still is an occult cause, a sort of archeus to which we refer all the facts which we do not understand and which we thenceforth consider explained, though they are often only classified and often wrongly classified: in the same way we attribute to catalytic [10] force any number of extremely obscure phenomena which, in my opinion, become more so when we refer them *en bloc* to an entirely unknown cause. In giving them the same name,

* H. Sainte-Claire Deville, *Leçons sur la dissociation prononcées devant la Société chimique.* Paris, 1866.

[10] All this applies to those recently invented forces, to the forces of dissolution and diffusion, to crystalogenic force, to all the special attractive and repelling forces brought in to explain phenomena of heating and superfusion, electric phenomena, etc.

we certainly believe we are placing them in the same category. But this classification is not even proved legitimate. What, indeed, could be more arbitrary than setting side by side catalytic phenomena that depend on the action or the presence of platinum sponge and concentrated sulphuric acid, when the platinum and the acid are not, so to speak, parties to the operation? These phenomena will perhaps later be explained in an essentially different way, according to whether they are produced under the influence of a porous material like platinum sponge or under the influence of a highly energetic chemical agent like concentrated sulphuric acid.

"In our studies, we must therefore lay aside unknown forces, to which we have recourse only because we have not measured their effects. On the contrary, we should direct our attention to the observation and numerical determination of effects, which alone are within our range. By such work, we establish differences and analogies, and new light results from comparisons and measurements.

"Thus heat and affinity are constantly face to face in our chemical theories. Affinity completely escapes us, and yet we attribute combination to it, as the effect of an unknown cause. Let us simply study the physical circumstances which accompany a combination, and we shall see how many measurable phenomena, how many curious relations present themselves at every moment. Heat, they say, destroys affinity. Let us patiently study the decomposition of bodies under the influence of heat measured in quantity of work, heat or energy: we shall immediately see how fruitful this study is, and how independent of every unknown force, unknown even from the point of view of the kind of units to which we must refer it for precise or approximate measurement. Especially in this sense affinity, considered as a force, is an occult cause, unless it simply expresses a quality of matter. In that case, it would be used simply to designate the fact that such and such substances can or cannot be combined in such and such definite circumstances."

When a phenomenon takes place outside the living body and does not occur in the organism, that is not because an entity called *life* prevents the phenomenon from taking place, but because the necessary condition for the phenomenon is not met with inside, as it is outside, the body. Thus it has actually been said that life prevents fibrin from coagulating in living animals' blood vessels, though it does coagulate outside the vessels because life no longer affects it. It is nothing of the sort; certain physico-chemical conditions are needed to make fibrin coagulate; they are harder to produce, but may nevertheless be found in the living, and, as soon as they appear, fibrin coagulates as well inside as outside the organism. The life invoked here is therefore only a physical condition which does or does not exist. I have shown that sugar is produced in the liver more abundantly after death than during life; certain physiologists drew the conclusion that life influences the formation of sugar in the liver; they said that life hinders its formation and death favors it. One is surprised to hear such vitalistic opinions in our day and to see them supported by men who pique themselves on applying to physiology and medicine the accuracy of physical science. I shall later show that here again physical conditions are either present or absent, but nothing else is real; because again, at the base of all these explanations, only the conditions of phenomena are to be found.

To sum up, we must learn that the words we use to express phenomena whose cause we do not know are nothing in themselves; and that the moment we grant them any value in criticism or discussion, we abandon experience and fall into scholasticism. In discussing or explaining phenomena, we must be very careful never to abandon observation or put a word in place of a fact. Very often we even expose ourselves to attack, solely because we abandon facts and conclude with a word that goes beyond what we have observed. The following example will prove it clearly.

Second example.—When I made my investigations of pancreatic juice, I noted that this fluid includes a peculiar material, pancreatin, which has characteristics of both albumen and casein. It resembles albumen in being coagulated by heat, but, like casein, differs from it in being precipitated by sulphate of magnesia. Magendie had made experiments, before me, on pancreatic juice, and had said that, according to his tests, pancreatic juice is a fluid containing albumen, while, from my investigations, I concluded that pancreatic juice does not comprise albumen, but does contain pancreatin, which is a material distinct from albumen. I showed my experiments to Magendie, pointing out that we disagreed on the conclusion, but that we nevertheless agreed on the fact that pancreatic juice is coagulated by heat; only that other new characteristics that I had seen prevented my deciding on the presence of albumen. Magendie answered: "This difference between us comes from my having inferred more than I saw; if I had simply said: pancreatic juice is a liquid coagulated by heat, I should have remained within the facts, and should have been unassailable." This example, which I have always remembered, shows how little value we should ascribe to words apart from the facts they represent. Thus the word albumen means nothing in itself; it merely recalls characteristics and phenomena. By extending this example to medicine, we should see that the words, fever, inflammation, and the names of diseases in general have no meaning at all in themselves.

When we create a word to characterize a phenomenon, we then agree in general on the idea that we wish it to express and the precise meaning we are giving to it; but with the later progress of science the meaning of the word changes for some people, while for others the word remains in the language with its original meaning. The result is often such discord that men using the same word express very different ideas. Our language, in fact, is only approximate, and even in science it is so indefinite that if we lose sight of phenomena and cling to words, we are speedily outside of reality. We therefore only injure science by arguing in favor of a word which is now merely a source of error, because it no longer expresses the same idea for everyone. Let us therefore conclude that we must always cling to phenomena and see in words only expressions empty of meaning, if the phenomena they should represent are not definite, or if they are absent.

The mind by its very nature has systematic tendencies; that is why we try to seek agreement about words rather than things. For experimental criticism, this is a false direction which confuses questions and makes us believe in differences of opinion which generally exist only in our way of interpreting phenomena, instead of having some bearing on the existence of facts and on their real importance. Like everyone who has had the good fortune of bringing into science unexpected facts or new ideas, I have been and still am the object of much criticism. Up to this time, I have not answered my opponents, because I have always had investigations on hand so that time and opportunity have been lacking; but in the remainder of this work, occasion to study them will quite naturally present itself, and by applying the principles of experimental criticism suggested in earlier paragraphs, we shall easily weigh the criticisms in question. Meantime I shall merely say that it is essential to distinguish between two things in experimental criticism: experimental fact and its interpretation. Science requires us first of all to agree on fact, because that is the basis on which we must reason. As to interpretations and ideas, they may vary, and discussing them is an actual advantage, because such discussion leads us to make other investigations and to undertake new experiments. In physiology, we should therefore never lose sight of the principles of true scientific criticism nor mix them with personalities or artifice. Among the artifices of criticism, many do not concern us because they are extra-scientific; one of them, however, we must point out. It consists in considering in a piece of work only what is defective and open to attack, while neglecting or concealing what is valid and important. This is the method of false criticism. In science the word criticism is not a synonym for disparagement; criticising means looking for truth by separating the true from the false and distinguishing the good from the bad. While just to men of science, such criticism alone is profitable for science. And this we shall easily show in the particular examples which we shall mention.

CHAPTER III

INVESTIGATION AND CRITICISM AS APPLIED TO EXPERIMENTAL MEDICINE

METHODS of investigation and of scientific criticism cannot vary from one science to another nor, for that matter, in different parts of the same science. It will therefore be easy to show that the rules for physiological investigation, suggested in the last chapter, are absolutely the same as those which should be followed in pathology and therapeutics. Thus methods of investigation of the phenomena of life should be the same in normal as in pathological conditions. This seems to us fundamental in biological science.

I. PATHOLOGICAL AND THERAPEUTIC INVESTIGATION

As in physiology, so in pathology and in therapeutics, the starting point of scientific investigation is now a casual fact or one occurring by chance, now an hypothesis, i.e., an idea.

I have sometimes heard physicians express the opinion that medicine is not a science, because all our knowledge of practical medicine is empirical and born of chance, while scientific knowledge is deduced with certainty from theories or principles. There is an error here, to which I wish to call attention.

All human knowledge had to begin with casual observations. Man indeed could know things only after seeing them; and the first time, necessarily, he saw them by chance; then he came to conceive ideas about things, to compare old facts and to deduce from them new ones; in a word, after empirical observation, he was no longer led to find other facts by chance, but by induction.

Fundamentally, then, all the sciences began with empiricism, that is to say, observation or chance experience had to form the first period. But empiricism is not a permanent state in any science. In the complex sciences of humanity, empiricism will necessarily govern practice much longer than in simpler sciences. Medical practice to-day is empirical in most cases; but that does not mean that medicine will never escape from empiricism. The complexity of its phenomena will make it harder to escape; but that should make us redouble our efforts to enter the scientific path as soon as we can. In a word, empiricism is not the negation of science, as certain physicians seem to think; it is only its first stage. We must even add that empiricism never wholly disappears from any science. Sciences, in fact, are not lighted up in every portion at once; they develop only a little at a time. In parts of physics and chemistry, empiricism still persists. This is proved every day by chance discoveries, unforeseen by prevailing theories. I therefore conclude that we make discoveries in the sciences only because all are still partially obscure. In medicine more numerous discoveries are still to be made, because almost everywhere empiricism and obscurity prevail. So this very complicated science is proved further behind the times than others; but that is all.

New medical observations are generally made by chance; if a patient with a hitherto unknown affection is admitted to a hospital where a physician comes for consultation, surely the physician meets the patient by chance. But a botanist in the field happens on an unfamiliar plant in exactly the same way; and by chance also an astronomer catches sight of a planet, whose existence he did not know of, in the sky. In such circumstances, the physician's originality consists in seeing the fact that chance presents to him and in not letting it escape, and his only merit is accurate observation. I cannot here analyze the characteristics of good medical observation. Reporting instances of chance medical observations would be just as dull. Medical works teem with them; everybody knows them. I shall therefore limit myself to saying in general that, to make a good medical observation, it is not only necessary to have an observing mind, but also to be a physiologist. We shall the better interpret the various meanings of a morbid phenomenon, we shall assign it the proper value, and we shall certainly not fall into the difficulty, with which Sydenham [1] reproached certain physicians, of putting important phenomena of a disease on the same plane as insignificant and accidental facts, like the botanist who described caterpillar bites among the characteristics of a plant. Besides, we must bring to observation of a pathological phenomenon, i.e., a disease, exactly the same state of mind and the same rigor, as to observation of a physio-

[1] Sydenham, *Médecine pratique.* Preface, p. 12.

logical phenomenon. We must never go beyond facts and must be, as it were, photographers of nature.

But once made, every medical observation becomes the starting point, as in physiology, for ideas and hypotheses which experimental physicians go on to investigate through fresh observations of patients or by experiments on animals.

We said that, in making physiological investigations, it often happens that a fresh fact arises unsought; that also occurs in pathology. To prove it, I need only cite the recent case of Zenker, who,[2] in pursuing his investigations of certain muscular changes in typhoid fever, found trichinæ which he was not looking for.

Pathological investigation may also take for its starting point a theory, an hypothesis or a preconceived idea. We might easily give examples to prove that absurd ideas, in pathology as in physiology, may sometimes lead to useful discoveries, just as it would not be hard to find arguments to prove that even the best accredited theories should be regarded only as temporary, and not as absolute truths to which facts should be bent.

Therapeutic investigation conforms to exactly the same rules as physiological and pathological investigation. Everyone knows that the first promoter of therapeutic science was chance, and that only by chance were the effects of most medicines first observed. Physicians have also often been guided in their therapeutic attempts by ideas; and it must also be said that they were often the strangest and most absurd theories or ideas. I need only cite the theories of Paracelsus, who deduced the action of drugs from astrological influences, and recall the ideas of Porta, who assigned medicinal uses to plants, deduced from their resemblances to certain diseased organs; thus carrots cured jaundice; lung-wort, phthisis, etc.[3]

Summing up, we cannot establish any valid distinction between methods of investigation that should be applied in physiology, in pathology and in hygiene. The method of observation and experiment is still the same, unchangeable in its principles and offering only a few peculiarities in its application, according to the relative complexity of phenomena. We cannot, indeed, find any radical dif-

---

[2] See *Rapport des prix de médecine et de chirurgie pour 1864* (*Compt. rendus de l'Acad. des sciences*).

[3] See Chevreul, *Considérations sur l'histoire de la partie de la médecine qui concerne la prescription des remèdes* (*Journal des savants*, 1865).

ference in the nature of physiological, pathological and therapeutic phenomena. Since all these phenomena depend on laws peculiar to living matter, they are identical in essence and vary only with the various conditions in which phenomena appear. We shall see later that physiological laws are repeated in pathological phenomena, whence it follows that the foundations of therapeutics must reside in knowledge of the physiological action of morbid causes, of medicines and of poisons; and that is just the same thing.

II. Experimental Criticism in Pathology and Therapeutics

Criticism of facts gives sciences their true individuality. All scientific criticism should explain facts rationally. If criticism is attributed, on the other hand, to personal feeling, science disappears; because such criticism rests on a criterion that can neither be proved nor conveyed as scientific truths should be. I have often heard physicians answer, when asked the reason for a diagnosis, "I do not know how I recognize such and such a case, but it is evident"; or when one asks them why they give certain remedies, they answer that they cannot exactly tell, and besides that they need not explain, since they are guided by their medical tact and intuition. It is easy to understand that physicians who reason in that way deny science. But we cannot too strongly protest against such ideas, which are bad, not only because they stifle every germ of science, but also because they especially encourage laziness, ignorance and charlatanism. I entirely understand a physician's saying that he cannot always rationally account for what he is doing, and I accept his conclusion that medical science is still plunged in the shades of empiricism; but if he goes on to proclaim his medical tact or his intuition as a criterion which he then means to impose on others without further proof, that is wholly antiscientific.

As in physiology, the only scientific criticism possible in pathology and in therapeutics is experimental criticism; and whether applied to ourselves or to the work of others, this criticism should always be based on absolute determination of facts. Experimental criticism, as we have seen, should reject statistics as a foundation for experimental therapeutic and pathological science. In pathology and therapeutics, we should repudiate undetermined facts, that is to say, those badly made, and sometimes imaginary, observations which are constantly brought forward as perpetual objections. As in physiology, there are crude facts which can enter into scientific reasoning only on condition that they be determined and exactly defined as to their necessary conditions.

But it is characteristic of criticism in pathology and therapeutics, first and foremost to require comparative observation and experiment. How, indeed, can a physician judge the etiology, if he does not make a comparative experiment to eliminate all the secondary circumstances, that might become sources of error, and make him take mere coincidences for relations of cause and effect? Especially in therapeutics, the need of comparative experiment has always struck physicians endowed with the scientific spirit. We cannot judge the influence of a remedy on the course and outcome of a disease if we do not previously know the natural course and outcome of the disease. That is why Pinel said in his clinic: "This year we will observe diseases without treating them, and next year we will treat them." Scientifically, we ought to adopt Pinel's idea without, however, accepting the long-range, comparative experiment which he proposed. Diseases, in fact, may vary in seriousness from one year to another; Sydenham's observations on the undetermined or unknown influence of what he calls the epidemic genius prove it. To be valid, comparative experiments have therefore to be made 'at the same time and on as comparable patients as possible. In spite of that, such comparisons still bristle with immense difficulties which physicians must strive to lessen; for comparative experiment is the *sine qua non* of scientific experimental medicine; without it a physician walks at random and becomes the plaything of endless illusions. A physician, who tries a remedy and cures his patients, is inclined to believe that the cure is due to his treatment. Physicians often pride themselves on curing all their patients with a remedy that they use. But the first thing to ask them is whether they have tried doing nothing, i.e., not treating other patients; for how can they otherwise know whether the remedy or nature cured them? Gall wrote a little known book [4] on the question as to what is nature's share and what is the share of medicine in healing disease, and he very naturally concludes that their respective shares are quite hard to

[4] Gall, *Philosophische medicinische Untersuchungen über Kunst und Natur im gesunden und kranken Zustand des Menschen.* Leipzig, 1800.

assign. We may be subject daily to the greatest illusions about the value of treatment, if we do not have recourse to comparative experiment. I shall recall only one recent example concerning the treatment of pneumonia. Comparative experiment showed, in fact, that treatment of pneumonia by bleeding, which was believed most efficacious, is a mere therapeutic illusion.[5]

From all this, I conclude that comparative observation and experiment are the only solid foundation for experimental medicine, and that physiology, pathology and therapeutics must be subject to this criticism in common.

[5] Béclard, *Rapport général sur les prix décernés en 1862* (*Mémoires de l'Académie de médecine*, Paris, 1863, Vol. XXVI, p. xxiii).

CHAPTER IV

PHILOSOPHIC OBSTACLES ENCOUNTERED BY
EXPERIMENTAL MEDICINE

ACCORDING to everything so far said in this Introduction, the principal obstacles encountered by experimental medicine lie in the enormous complexity of the phenomena studied. I need not return to this point already explained from every angle. But besides these wholly material and, so to speak, objective difficulties, there are obstacles to experimental medicine arising from vicious methods, bad mental habits and certain false ideas about which we shall now say a few words.

I. THE FALSE APPLICATION OF PHYSIOLOGY TO MEDICINE

I certainly do not claim to have been the first to propose applying physiology to medicine. That was long ago recommended, and numerous attempts have been made in this direction. In my works and my teaching at the Collège de France, I am therefore merely following out an idea which is already bearing fruit through its application to medicine. More than ever to-day, young physicians are advancing along this path, rightly considered the path of progress. However, I have frequently seen the application of physiology to medicine misunderstood, so that it not only fails to produce the good results which we have a right to expect, but becomes actually harmful, and thus furnishes arguments to the detractors of experimental medicine. It is therefore most important to make the subject plain; for in dealing with the important question of method, we shall find a fresh opportunity to define more exactly the true point of view of what we call experimental medicine.

Experimental medicine differs in object from the medicine of observation, just as the sciences of observation in general differ from the experimental sciences. The object of any science of observation is to discover the laws of natural phenomena so as to foresee them;

but it cannot master them or alter them at pleasure. Astronomy is typical of these sciences; we can foresee astronomical phenomena, but we cannot change them in any way. The object of an experimental science is to discover the laws of natural phenomena, for the purpose not only of foreseeing them, but of regulating them at pleasure and mastering them: such are physics and chemistry.

Among physicians, there are some who actually believe that medicine should remain a science of observation, i.e., that it should be able to foresee the course and outcome of diseases, but should not directly act on disease. There are others, and I am one of them, who think that medicine can be an experimental science, i.e., that it should delve into the interior of organisms and find ways of altering and, to a certain extent, regulating the hidden springs of living machines. Observing physicians look on a living organism as a little world contained in the great world, like a kind of ephemeral living planet whose motions are ruled by laws which we discover by simple observation, so as to foresee the progress and evolution of vital phenomena in health or disease, but without ever being able to alter their natural course in any way. This doctrine is found in Hippocrates in its purest form. Medicine of simple observation obviously excludes all manner of active medical intervention; for this reason it is also known as *expectant medicine,* that is to say, medicine that observes and foresees the course of diseases without aiming to act directly on their progress.[1] It is rarely that we find a physician purely Hippocratic in this respect, and it would be easy to prove that many physicians, who loudly applaud Hippocratism, do not trust to its precepts in the least when they give themselves up to the most active and disordered flights of empirical medication. Not that I condemn these therapeutic attempts which, most of the time, are only experimentations *to see;* only I say that this is not Hippocratic medicine, but empiricism. Empirical physicians, acting more or less blindly, are, after all, experimenting on vital phenomena, and thus class themselves in the empirical period of experimental medicine.

Experimental medicine is therefore medicine that claims knowledge of the laws of healthy and diseased organisms, not only so as to foresee phenomena, but also so as to be able to regulate and alter them

---

[1] *Leçon d'ouverture du cours de médecine au Collège de France (Revue des cours scientifiques,* Dec. 31, 1864).

within certain limits. Accordingly, we easily perceive that medicine
necessarily tends to become experimental, and that every physician
who gives his patients active medicines coöperates in building up
experimental medicine. But if such action, on the part of experimenting physicians, is to transcend empiricism and deserve the
name of science, it must be based on knowledge of the laws governing action in the organism's inner environment, whether in a healthy
or a pathological state. The scientific basis of experimental medicine is physiology; we have often said this; it must be proclaimed
aloud, because without it no medical science is possible. Diseases
at bottom are only physiological phenomena in new conditions still
to be determined; toxic and medicinal action, as we shall see, come
back to simple physiological changes in properties of the histological
units of our tissues. In a word, physiology must be constantly applied to medicine, if we are to understand and explain the mechanism
of disease and the action of toxic and medicinal agents. Now, precisely this application of physiology must here be carefully defined.

We saw above how experimental medicine differs from Hippocratism and from empiricism; but we did not say that experimental medicine should therefore renounce observational medicine
or the empirical use of medicines; far from it, experimental medicine makes use of medical observation as a necessary support. In
fact, experimental medicine never systematically rejects any fact
or popular observation; it must examine everything experimentally,
and it seeks the scientific explanation of facts which observational
medicine and empiricism have already noted. Experimental medicine, then, is what I might call the second period of scientific
medicine, the first period being observational medicine; and quite
naturally, therefore, the second period is added to the first and rests
on it. The first requirement, then, in practising experimental medicine, is to be an observing physician and to start from pure and
simple observations of patients made as completely as possible; experimental science comes next, analyzing every symptom by trying
to connect it with explanations and vital laws that shall include
the relation of the pathological state to the normal or physiological
condition.

But in the present state of biological science, no one can presume
to explain pathology by physiology alone; we must move in that
direction because it is the scientific path, but we must shun belief

in the illusion that our problem is solved. For the moment, therefore, the prudent and reasonable thing to do is to explain all that we can explain in a disease by physiology and leave what is still inexplicable to the future progress of biological science. This kind of analysis, advancing only step by step as the progress of physiological science permits, isolates the essential elements of a disease by elimination, a little at a time, grasps its characteristics more accurately and allows us to guide therapeutics more intelligently. Besides, the analytic, progressive advance still keeps the individual character and aspect of the disease. But if we take advantage, instead, of a few possible connections between pathology and physiology, to try to explain the whole disease at a single stroke, then we lose sight of the patient, we distort the disease, and by our false application of physiology, we retard experimental medicine, instead of promoting its progress.

Unfortunately I must blame not only pure physiologists for the wrong application of physiology to pathology, but also professional pathologists and physicians. In various recent publications on medicine, whose physiological tendencies, by the way, I approve and praise, I see that before any exposition of medical observations, the authors begin with a summary of everything learned by experimental physiology about phenomena connected with the disease with which they are concerned. Then they contribute observations of patients, sometimes without definite scientific object, sometimes to show that physiology and pathology are in agreement. But aside from the fact that agreement is not always easy to prove, because points in experimental physiology are often still under consideration, I find this sort of procedure essentially disastrous to medical science, in that it subordinates the more complex science, pathology, to physiology, a simpler science. This is, in fact, the inverse of what we previously said should be done: we should first of all state the medical problem as given by observation of the disease, then try to find the physiological explanation, by experimentally analyzing the pathological phenomena. But in this analysis, medical observation must never disappear or be lost sight of; it must remain as the constant basis or common ground of all our studies and explanations.

In this work, I cannot develop as a whole the things that I have just said, because I have had to limit myself to giving the results of my experience in physiological science with which I am most familiar. In publishing a simple essay on the principles of scientific medicine, my idea is to be of some use to medicine. Medicine, indeed, is so vast that we can never hope to find a man able to cultivate all parts of it fruitfully at one time. But in the part where each physician takes up his quarters, he must thoroughly understand the scientific connections between all the medical sciences, so as to avoid scientific anarchy by guiding his investigations in a direction useful to the whole. I am not practising clinical medicine here; but I must take account of it, nevertheless, and assign it the first place in experimental medicine. So if I were planning a treatise on experimental medicine, I should go to work by invariably making observation of disease the basis of every experimental analysis. I should then proceed with my explanations, symptom by symptom, until I had exhausted all the information obtainable from present experimental physiology, and the result of all this would be medical observation reduced to its simplest terms.

In saying above that we must explain by experimental physiology only what can be explained in disease, I do not want my idea misunderstood or taken as an admission that there are things in disease which can never be physiologically explained. My idea is just the reverse, because I believe that we shall explain everything in pathology, but little by little and in step with the development of experimental physiology. We can just now explain nothing about certain diseases, for instance the eruptive diseases, because the related physiological phenomena are unknown. So the objection, which some physicians find here, to physiology as a help to medicine, is not worthy of consideration. That kind of argumentation is tinged with scholasticism and proves that those who use it have no correct idea of such a science as experimental medicine can be.

To sum up, as the natural foundation of experimental medicine, experimental physiology cannot suppress observation of the sick or lessen its importance. Moreover, physiological knowledge is not only indispensable in explaining disease, but is also necessary to good clinical observation. For example, I have seen observers surprised into describing as accidents certain thermal phenomena which occasionally result from nerve lesions; if they had been physiologists, they would have known how to evaluate morbid symptoms which are really nothing but physiological phenomena.

II. Scientific Ignorance and Certain Illusions of the Scientific Spirit Hinder the Development of Experimental Medicine

We have just said that knowledge of physiology is indispensable to physicians; we must therefore cultivate the physiological sciences, if we wish to further the development of experimental medicine. This is all the more necessary, because it is the only way to provide a foundation for scientific medicine, and unfortunately we are still far from the time when we shall see the scientific spirit generally prevailing among physicians. Now the absence of the scientific habit of mind is a serious hindrance, because it favors belief in occult forces, rejects determinism in vital phenomena, and leads to the notion that the phenomena of living beings are governed by mysterious, vital forces which are continually invoked. When an obscure or inexplicable phenomenon presents itself, instead of saying "I do not know," as every scientific man should do, physicians are in the habit of saying, "This is life"; apparently without the least idea that they are explaining darkness by still greater darkness. We must therefore get used to the idea that science implies merely determining the conditions of phenomena; and we must always seek to exclude life entirely from our explanations of physiological phenomena as a whole. Life is nothing but a word which means ignorance, and when we characterize a phenomenon as vital, it amounts to saying that we do not know its immediate cause or its conditions. Science should always explain obscurity and complexity by clearer and simpler ideas. Now since nothing is more obscure, life can never explain anything. I emphasize this point, because I have seen even chemists at times appeal to life to explain certain physico-chemical phenomena peculiar to living beings. Thus the ferment in yeast is an organic, living material which has the property of converting sugar into alcohol, carbonic acid and several other products. I have sometimes heard it said that the property of decomposing sugar was due to the life inherent in a globule of yeast. This vitalistic explanation means nothing and explains nothing about the action of yeast. We do not know the nature of this property, but it must necessarily belong to the physico-chemical order and be as precisely defined as, for instance, the property of platinum sponge which produces a more or less analogous action that cannot be attributed to vital force. In a word, all the properties of living matter are, at bottom, either known and defined properties, in which case we call them physico-chemical properties, or else unknown and undefined properties, in which case we name them vital properties. Certainly a special force in living beings, not met with elsewhere, presides over their organization; but the existence of this force cannot in any way change our idea of the properties of organic matter,—matter which, when once created, is endowed with fixed and determinate, physico-chemical properties. Vital force is, therefore, an organizing and nutritive force; but it does not in any way determine the manifestation of the properties of living matter. In a word, physiologists and physicians must seek to reduce vital properties to physico-chemical properties, and not physico-chemical properties to vital properties.

The habit of vitalistic explanation makes us credulous and promotes the introduction of erroneous or absurd data into science. Thus, quite recently I was consulted by an honorable and much respected practising physician who asked my opinion of a most unusual case, of which, he said, he was very sure, because he had taken all precautions necessary to observing it well: here was a woman in good health except for a few nervous anomalies, who had neither eaten nor drunk anything for several years. Evidently the physician was persuaded that vital force is capable of anything, so that he sought no other explanation. The slightest idea of science, however, and the simplest notions of physiology, would have been enough to undeceive him, by showing that his statement very nearly amounted to saying that a candle can go on shining and burning for several years without growing any shorter.

Belief that the phenomena of living beings are dominated by an indeterminate vital force often also gives experimentation a false basis and puts a vague word in place of exact experimental analysis. I have seen physicians submit questions to experimental analysis in which they took as their starting point the vitality of certain organs, the idiosyncrasy of certain individuals or the antagonism of certain medicines. Now, vitally, idiosyncrasy and antagonism are merely vague words which should first be qualified and reduced to a definite meaning. In the experimental method, then, it is a matter of absolute principle always to take, as our starting point for experimentation or reasoning, an exact fact or a good observation, and not a vague word. When the discussions of physicians and naturalists lead to nothing, it is usually because they fail to conform to this analytic precept. In a word, in experimentation on living beings, as with inorganic bodies, it is essential, before beginning our experimental analysis of a phenomenon, to make sure that the phenomenon exists and never to let ourselves be deceived by words that lose sight of facts as they are.

As we have elsewhere explained, doubt is the foundation of experimentation; yet we must not confuse philosophic doubt with that systematic negation which casts doubt on the very principles of science. We must doubt only theories, and we must doubt even them only to the point of experimental determinism. Some physicians believe that the scientific spirit sets no limit to doubt. Aside from these physicians, who deny medical science by admitting that nothing positive can be known, others deny it by the opposite method, admitting, as they do, that they have learned their medicine they know not how, and are masters of it through a kind of intuitive science which they call clinical sense or instinct. In medicine, as in other practical sciences, I do not of course question the existence of what is called tact or clear-sightedness. Everyone knows, in fact, that habit may give a kind of empirical knowledge of things sufficient to guide practitioners, even though they cannot always precisely account for it at first. But what I blame is willfully staying in this empirical state and not trying to get out of it. By attentive observation and study, we can always manage to account for our actions and so succeed in transmitting our knowledge to others. Besides, I do not deny that the practice of medicine has severe requirements; but here I am talking pure science and am attacking medical tact as an anti-scientific datum whose natural exaggerations are decidedly harmful to science.

Another false opinion, which is pretty well accredited and even professed by great practising physicians, is expressed in saying that medicine is not destined to become a science, but only an art, and that physicians accordingly should be artists, not men of science. I find this idea erroneous and essentially harmful to the development of experimental medicine. First, what is an artist? An artist is a man who carries out a personal idea or feeling in a work of art. Here, then, are two things: the artist and his work; the artist is necessarily judged by his work. But what can a medical artist be? If he is a physician who treats disease according to his personal idea or feeling, then where is the work of art by which the medical

artist is to be judged? Is it cure of the disease? That would be
a strange kind of work of art, and the physician's authorship would
be seriously disputed by nature. When a great painter or a great
sculptor makes a beautiful picture or a magnificent statue, no one
imagines that the statue grew out of the earth or that the picture
made itself, while we can perfectly well maintain that a disease has
cured itself and can often prove that the cure would have been
better without the artist's interference. Then what has become of
the criterion, or medical work of art? The criterion evidently disappears; because no physician's ability can be judged by the number of patients that he says he has cured; he must first of all prove
scientifically that it was he who cured them, and not nature. I
shall not further emphasize this untenable medical claim to art.
In reason, physicians can be men of science only, or, in the meantime, empiricists. Empiricism, which means experience at bottom
(ἐμπειρία, experience), is only unconscious or non-rational experience, acquired by every-day observation of facts, in which the experimental method itself originates (see p. 12). But as we shall see
again in the next paragraph, empiricism in its true sense is merely
the first step in experimental medicine. Empirical physicians
should strive toward science, for though they often decide in practice
according to unconscious experience, they should at least still guide
themselves by induction based on as solid medical learning as possible. In a word, since there is no such thing as a medical work of
art, there is no such thing as a medical artist; physicians calling themselves such injure medical science, because they exalt a physician's
personality by lowering the importance of science; thus they prevent
men from seeking, in the experimental study of phenomena, the
support and criterion which they believe they, through inspiration
or mere feeling, have within themselves. But as I just said, this
supposed therapeutic inspiration is often supported by no other
proofs than some chance fact which might favor an untaught man
or a charlatan, just as much as an educated man. This bears no sort
of relation to the artist's inspiration which is embodied at last in a
work judged by all the world, and always requiring, for its execu-
tion, exact study often accompanied by unwearied labor. In my
opinion, then, the inspiration of physicians, who do not rely on
experimental science, is mere fantasy; and in the name of science
and humanity they should be rebuked and proscribed.

To sum up, experimental medicine, which is a synonym for scientific medicine, can be established only by spreading the scientific spirit more and more among physicians. In my opinion, the one thing to do, to reach this goal, is to give our young men solid instruction in experimental physiology. I do not mean to say that physiology is the whole of medicine; I have explained myself elsewhere on this point, but I do mean to say that experimental physiology is the most scientific part of medicine, and that in studying it, young physicians will acquire scientific habits which they will later carry into pathological and therapeutic investigation. The wish that I am expressing here roughly corresponds to Laplace's idea: when he was asked why, since medicine was not a science, he had proposed admitting physicians to the Academy of Sciences; he answered. "This is why: to get them among men of science."

III. EMPIRICAL AND EXPERIMENTAL MEDICINE ARE BY NO MEANS INCOMPATIBLE; ON THE CONTRARY, THEY MUST BE INSEPARABLE

For a long time, men have said and repeated that the physicians most learned in physiology are the worst physicians, and that they are the most awkward when action is necessary at the patient's bedside. Does this mean that physiological science is harmful to practice? In that case, I must have taken a completely false point of view. We must therefore carefully study this opinion, which is a favorite theme of many practising physicians, but which I, for my part, consider completely erroneous.

To begin with, we must remember that the practice of medicine is exceedingly complex, involving any number of social and extra-scientific questions. Even in practical veterinary medicine, therapeutics is often dominated by considerations of profit or of agriculture. I recall my membership in a commission studying what was to be done to prevent the ravages of certain murrains of horned cattle. We were all weighing physiological and pathological considerations, to decide on the proper treatment to cure the sick animals, when a practising veterinarian took the floor to say that this was not the question; and he proved clearly that curative treatment would be the ruin of agriculture, and that the best thing to do was to slaughter the sick animals and turn them to the best possible account. Considerations of this kind never enter into human medicine, because preserving human life is the physician's sole aim. Yet physicians, in their treatment, often have to take account of the so-called influence of the moral over the physical, and also of any number of family and social considerations which have nothing to do with science. Therefore, an accomplished practising physician should be not only learned in his science, but also upright and endowed with keenness, tact and good sense. Practising physicians exert an influence in every rank of society. In numberless cases, physicians are the custodians of state interests in major affairs of public administration; at the same time they are the confidants of families and often hold reputation and most cherished interests in their hands. Able practitioners can acquire great and legitimate influence among men, because apart from science, they have a moral influence on society. And so, like Hippocrates, everyone having the dignity of medicine at heart has always insisted strongly on moral qualities in physicians.

I have no intention of discussing here the social and moral influence of physicians nor of penetrating what might be called the mysteries of medical practice; I am simply treating the scientific side and am separating it so as to judge its influence better. I certainly do not here intend to study the question whether an educated physician would treat his patients better or worse than an uneducated one. Put in that form, the question would be absurd; I naturally assume two physicians equally well educated in methods of treatment, and I intend to consider here only whether the *scientific* physician, i.e., the physician endowed with the experimental spirit, will treat his patient less successfully than the *empirical* physician who contents himself with noting facts solely on the basis of medical tradition, or the systematic physician who acts according to the principles of some doctrine or other.

In medicine, there have always been two divergent tendencies resulting from the very nature of things. The first tendency in medicine, arising from the kindly feelings of man, is to help a neighbor in trouble, and to relieve him with remedies or by moral or religious means. Medicine must therefore have been mingled with religion, from its beginning, while possessing at the same time numberless more or less active agents. Found by chance or of necessity, these remedies were later handed down by tradition, either alone or together with religious practices. But after this first flight, which started, so to speak, from the heart, men must have begun to reflect, and seeing the sick recover of themselves and without medicine, they were inclined to ask, not only whether the medicines given were useful, but whether they were not harmful. The first medical reflection, or first medical reasoning, resulting from study of the sick, made men recognize a spontaneous, medicinal force in the living organism; and observation taught them to respect it and try merely to guide and help it in its fortunate tendencies. The first steps in scientific medicine taken by Hippocrates involved a doubt about the curative results of empirical methods and the appeal to the laws of living organisms to effect the cure of the sick. But this kind of medicine, founded as science on observation, and as treatment on expectancy, still allows other doubts to subsist. While recognizing how direful for the patient it may be to use empirical medicaments to disturb the tendencies of nature when they are favorable, men must have asked themselves, on the other hand, whether it might not be possible, and useful to the patient, to disturb and change them when they were bad. It was therefore no longer merely a case of physicians guiding and helping nature in its fortunate tendencies; *Quo vergit natura, eo ducendum,* but also of combating and dominating nature in its evil tendencies, *medicus naturae superator.* The heroic remedies, the universal panaceas, the specifics of Paracelsus and others, are merely the empirical expression of a reaction against Hippocratic medicine, i.e., against expectancy.

By its very nature, experimental medicine has no system and rejects nothing in the way of treatment or cure of disease; it believes and accepts everything that is founded on observation and proved by experience. Though we have repeated it often already, we must here recall the fact that experimental medicine, as it is called, is certainly not a new theory of medicine. It is one with the medicine of all people and times, in all its solid gains and sound observations. Scientific, experimental medicine goes as far as possible in the study of vital phenomena; it cannot limit itself to observing diseases or content itself with expectancy or stop at remedies empirically given, but in addition it must study experimentally the mechanism of diseases and the action of remedies, so as to account for them scientifically. Above all, the analytic spirit of the experimental method in modern science must be brought into medicine; but this will not absolve experimental physicians from being good observers; they must be thoroughly educated in clinics, must know diseases accurately in all their normal, abnormal and insidious forms, be familiar with every method of pathological investigation, and be good, as we say, in diagnosis and prognosis. Besides this, they must be consummate therapeutists and know everything that empirical or systematic attempts have taught us about the action of remedies in different diseases. In a word, experimental physicians, like all educated physicians, must have every kind of knowledge that we have just enumerated; but they will differ from systematic physicians in not conducting themselves according to any system; but, instead of taking as their goal observation of disease and notation of the action of remedies, they will be distinguished from Hippocratic and empirical physicians by their will to go further and, with the help of experimentation, enter into the explanation of vital mechanisms. For their part, Hippocratic physicians are satisfied when they succeed in clearly describing a disease in its course, in learning and foreseeing its various favorable or direful endings by exact signs, so as to be able to intervene, if necessary, to help nature and to guide it toward a happy ending; scientific medicine, they believe, should set itself this goal. Empirical physicians are satisfied when, with the help of empiricism, they succeed in knowing that a given remedy cures a given disease, in learning the exact doses in which to administer it, and the cases in which it must be used; they may also believe they have reached the limits of medical science. But while experimenting physicians are the first to admit and understand the scientific and practical importance of the preceding ideas, without which medicine could not exist, they do not believe that medicine as a science should stop at observation and empirical knowledge of phenomena or be satisfied with somewhat vague systems. So that Hippocratic, empirical and experimenting physicians do not differ in the least in the nature of their knowledge; they differ only in the mental point of view which leads them to carry the medical problem somewhat further. The mediating power of nature invoked by Hippocratists, and the therapeutic or other force assumed by empiricists are simple hypotheses in the eyes of experimenting physicians. With the help of experimentation, they must penetrate into the inmost phenomena of living machines and define their mechanism in its normal as well as its pathological state. We must investigate the immediate causes of normal phenomena, which should be found in definite organic conditions in relation to the properties of fluids and tissues. It is not enough to know the phenomena of mineral nature empirically as well as their results; but physicists and chemists mean to go back to their necessary conditions, i.e., to their immediate causes, so as to be able to regulate their manifestation. In the same way, it is not enough for physiologists to know empirically the normal and abnormal phenomena of living nature; but like physicists and chemists, they mean to go back to the immediate causes of phenomena, i.e., to their necessary conditions. In a word, it is not enough for experimenting physicians to know that quinine cures fever; but what is above all significant to them is knowing what fever is and accounting for the mechanism by which quinine cures. All this is significant to experimenting physicians because, as soon as they know it, the fact of curing fever with quinine will no longer be an empirical, isolated fact, but a scientific fact. This fact will then connect itself with conditions which will relate it to other phenomena, and we shall thus be led to knowledge of the laws of organisms and the possibility of regulating their manifestations. Experimental physicians are therefore concerned most of all with seeking to establish medical science on the same principles as all the other experimental sciences. Let us now see how a man animated with this scientific spirit should behave at a patient's bedside.

Hippocratists, believing in a mediating nature, and but little in the curative effect of drugs, quietly follow the course of a disease; in almost passive expectancy, they limit themselves to encouraging the fortunate tendencies of nature with a few simple medicines. Empiricists, with their faith in the efficacy of drugs as a means of changing the direction of diseases and curing them, content themselves with empirically noting medicinal effects, without trying to understand their mechanism scientifically. They are never perplexed: when one remedy fails, they try another; they always have receipts or formulæ at hand for any and every case, because they draw on an immense therapeutic arsenal. Empirical medicine is certainly the most popular. People believe that through a kind of compensation nature provides a remedy for every ill, and that medicine consists in a collection of recipes for all ills, handed down to us, age by age, since the beginnings of the healing art. Experimenting physicians are Hippocratists and empiricists at one and the same time, in that they believe in the power of nature and the efficacy of drugs; only they want to know what they are doing; it is not enough for them to observe and to act empirically, they want to experiment scientifically and to understand the physiological mechanism producing disease and the medicinal mechanism effecting a cure. If this were their exclusive mental tendency, it is true that experimenting physicians would be as much, as empirical physicians are little, perplexed. Indeed in the present state of medicine, we know so little about the action of drugs that, if experimenting physicians were logical, they would be reduced to doing nothing and to remaining most of the time in the state of expectancy enjoined by their doubts and uncertainties. In this sense it is possible to say that scientific physicians are always the most perplexed at a patient's bedside. That is thoroughly true; they are really perplexed, because, on the one hand, they are convinced that we can take action with the help of powerful medicinal means, while, on the other hand, their ignorance of the mechanism of such action holds them back, for the experimental scientific spirit is utterly averse to producing effects and studying phenomena without trying to understand them.

There is evidently an excess of these two radical turns of mind among empiricists and among experimenters: in practice the two points of view should be fused, and the seeming contradiction between them should disappear. What I am saying here is by no means a kind of compromise or arrangement for convenience in medical practice. I am maintaining a purely scientific opinion, because I can easily prove that the true experimental method consists in a logical union of empiricism and experimentation. In fact, we have seen that, before foreseeing facts according to the laws which govern them, we must first observe them empirically or by chance; just as before experimenting along the lines of a scientific theory, we must first experiment empirically, in order to see. Now, in this respect, empiricism is nothing but the first step of the experimental method; for, as we said, empiricism cannot be a final stage; the vague, unconscious experience, which may be called medical tact, is later transformed into a scientific idea, by the experimental method which is conscious and logical. Experimental physicians, therefore, are empirical to begin with; but instead of stopping at that, they try to pass through empiricism so as to reach the second step in the experimental method, i.e., exact and conscious experiment which gives experimental knowledge of the law of phenomena. In a word, we must suffer empiricism; but trying to set it up as a system is an antiscientific tendency. As for systematic and doctrinal physicians, they are empiricists who, instead of having recourse to experimentation, take pure hypotheses, or else the facts taught them by empiricism, and join them together with the help of an ideal system, from which they later deduce their line of medical conduct.

Consequently, I think that experimenting physicians who wish to use, at the patient's bedside, only medicines whose physiological effect they understand would exaggerate in a direction that made them distort the true meaning of the experimental method. Before understanding facts, experimenters must first note them and free them from every source of error with which their minds be tainted. Experimenters must therefore first apply their minds to collecting medical or therapeutic observations empirically made. But they do still more; they are not limited to subjecting to the experimental criterion all the empirical facts that medicine presents to them; they go out to meet them. Instead of waiting for chance or accidents to teach them the effects of medicines, they try them empirically on animals, to get indications to guide them in the experiments that they afterward make on man.

I consider then that true experimenting physicians should be no more perplexed at a patient's bedside than empirical physicians. They will make use of all the therapeutic means advised by empiricism; only instead of using them according to authority and with a confidence akin to superstition, they will administer them with that philosophic doubt which is appropriate to true experimenters; they will verify the results on animals, and by comparative observations on man, so as to determine rigorously the relative influence of nature and of medicine in curing disease. In case it is proved that the remedy does not cure, and all the more so if it is shown to be harmful, experimenters should renounce it, and, like the Hippocratists, should await events. Certain practising physicians, fanatically convinced of the excellence of their medications, cannot understand the experimental therapeutic criticism of which I have just spoken. They say we can give sick people only medicines in which we have faith, and they think administering to our neighbors a remedy, which we doubt, is a failure of medical morals. I do not accept this reasoning, for it would lead us both to deceive ourselves and to deceive others without scruple. As for myself, I think it better to try to enlighten ourselves, so as to deceive no one.

Experimenting physicians should, therefore, not be mere physiologists waiting with folded arms for experimental medicine to be established scientifically, before taking action in behalf of their patients. Far from it, they should use the remedies that are empirically known, not only on equal terms with empiricists, but should go even further and try new medicines according to the rules suggested above. Experimental physicians, then, like empiricists, should be able to aid the sick by every means in the possession of practical medicine. What is more, with the help of the scientific spirit that guides them, they will do their part in founding experimental medicine; and that should be the most ardent wish of all physicians who want to see medicine rise out of its present state. We must suffer empiricism, as we said, as a transient and imperfect stage of medicine, but must not set it up as a system. In our faculties of medicine, we must therefore not limit ourselves, as men have actually said, to making empirical healers; that would degrade medicine and reduce it to the level of business. First of all, we must inspire young men with the scientific spirit and initiate them into the ideas and tendencies of modern science. Doing anything else would be inconsistent, besides, with the great variety of knowledge demanded of doctors, solely to enable them to cultivate medical science; for much narrower knowledge is demanded of health officers who are concerned only with empirical practice.

But the objection may be raised that experimental medicine, about which I am talking at such length, is a theoretic conception whose reality has not yet been vindicated in practice; because facts have not demonstrated that we may expect the same scientific precision in medicine as in the experimental sciences. As far as possible, I wish to leave no doubt in the reader's mind and no ambiguity in my own thought; I am therefore going to return to this subject with a few words, in order to show that experimental medicine is only the natural blossom of practical medical investigation, guided by a scientific spirit.

I said above that compassion and blind empiricism are the prime movers of medicine; later came reflection bringing doubt, then scientific verification. This medical evolution can still be verified around us every day, for every man goes on learning, as does all humankind.

Expectancy, whatever help it may give the tendencies of nature, can be only an incomplete method of treatment. Moreover, we must often act against the tendencies of nature. If, for example, an artery is open, we clearly must not favor nature which makes the blood come out and leads to death. We must act in the opposite direction, stop the hemorrhage and save a life. Just so, when a patient has an attack of septicemia, we must act against nature and stop the fever if we mean to cure our patient. Empiricists, then, may save patients whom expectancy would leave to die, just as expectancy might permit the recovery of a patient whom empiricism would kill. So that empiricism is also an insufficient method of treatment, in that it is uncertain and often dangerous. Now experimental medicine is only a union of expectancy with empiricism, enlightened by reasoning and experimentation. But experimental medicine will be the last to establish itself, and only then can medicine become scientific. We shall see, in fact, that all parts of medical knowledge are interrelated and are necessarily subordinate one to another in their evolution.

When a physician is called to a patient, he should decide on the diagnosis, then the prognosis, and then the treatment of the disease. The diagnosis can be established only through observation; in recognizing a disease, physicians connect it with some form of disease already observed, known and described. Observation also gives the progress and prognosis of the disease; physicians must know the evolution of the disease, its duration and gravity in order to predict its course and outcome. Here statistics intervene to guide physicians, by teaching them the proportion of mortal cases; and if observation has also shown that the successful and unsuccessful cases can be recognized by certain signs, then the prognosis is more certain. Finally comes the treatment: when physicians are Hippocratists, they limit themselves to expectancy; when they are empiricists they give remedies, still basing their action on observation which has taught by experiments or otherwise that such and such a remedy has succeeded in this disease a certain number of times; when physicians are systematic, they may accompany their treatment with vitalistic or other explanations, that will make no difference in the result. Here again, only statistics are invoked to establish the value of the treatment.

Such, in fact, is the state of empirical medicine, which is conjectural medicine because it is based on statistics which collect and compare cases that are analogous or more or less similar in their outer characteristics, but undefined as to their immediate causes.

Conjectural medicine must necessarily precede exact medicine, which I call experimental medicine because it is based on the experimental determination of the cause of disease. In the meantime, we must resign ourselves to practising conjectural or empirical medicine; but, I repeat, though I have often said it before, we must recognize that medicine should not stop there, and that it is destined to become experimental and scientific. We are doubtless far from the time when all medicine will be scientific; but that need not prevent our conceiving it possible and making every effort to strive toward it, by trying even to-day to introduce into medicine the method that must lead us to that goal.

Medicine will necessarily first become experimental in the diseases most easy of experimental approach. Among these, let me choose an example to show my idea of how empirical medicine can become scientific. The itch is a disease whose causation is now pretty well defined scientifically; but this has not always been the case. Formerly we knew the itch and its treatment only empirically. Then we guessed about lesions in the itch and collected statistics on the value of one salve or another for curing the disease. Now that the cause of the itch is known and experimentally determined, it has all become scientific, and empiricism has disappeared. We know the tick, and by it we explain the transmission of the itch, the skin changes and the cure, which is only the tick's death through appropriate application of toxic agents. No further hypotheses need now be made about the metastasis of the itch, no further statistics collected about its treatment. We cure it *always* without any exception, when we place ourselves in the known experimental conditions for reaching this goal.[2]

Here, then, is a disease that has reached the experimental stage; and physicians are masters of it just as much as physicists and chem-

[2] Hardy, *Bulletin de l'Académie de médecine.* Paris, 1863-64, Vol. XXIX, p. 546.

ists are masters of a phenomenon of mineral nature. Experimenting physicians will exert their influence successively on diseases one by one, as soon as they experimentally learn their correct determinism, i.e., their immediate cause. Even the best informed empirical physicians lack the experimenter's sureness. One of the clearest cases of empirical treatment is curing fever with quinine; yet this cure is far from being as certain as curing the itch. Diseases that have their seat in the outer organic environment, such as epidemic diseases, are the easiest to study and to analyze experimentally. These diseases will more quickly reach the stage where their causation is known and their treatment scientific. But later, in proportion as physiology progresses, we shall be able to get at the inner environment, i.e., the blood, discover there the parasitic and other changes that cause diseases and determine the medicinal, physico-chemical or specific agents capable of acting in this inner environment, altering the pathological mechanisms located there and reëchoing thence throughout the whole organism.

My conception of experimental medicine is summed up above. As I have often repeated, it is nothing but the consequence of the wholly natural evolution of scientific medicine. In this respect, medicine does not differ from other sciences which have all passed through empiricism before reaching their final experimental stage. In chemistry and in physics, practical methods of extracting metals, making magnifying glasses, etc., were known before the scientific theory evolved.

Empiricism, then, also guided these sciences through their nebulous days; but only since the advent of experimental theories have physics and chemistry taken such brilliant flights as applied sciences, for we must be careful to avoid confusing empiricism with applied science. Applied science always implies pure science as its support. Medicine will doubtless pass through empiricism much more slowly and laboriously than the physico-chemical sciences; not only because the organic phenomena with which it is concerned are much more complex, but also because the requirements of medical practice, which I need not study here, help to keep medicine in the personal realm, and thus oppose the experimental development. I need not here return to what I have elsewhere so amply explained, to wit, that the spontaneity of living beings does not prevent the application of the experimental method, and that knowledge of the simple or complex causation of vital phenomena is the one foundation of scientific medicine.

The object of experimenting physicians is to discover and grasp the original causation of a series of obscure and complex morbid phenomena; as a result they will dominate all secondary phenomena; thus we have seen that, on mastering the tick which causes the itch, we naturally master all the derived phenomena. By learning the ultimate cause of poisoning with curare, we easily explain all secondary phenomena; and to find a cure, we must always go back, in the end, to the original causation of phenomena.

Medicine is destined, then, to get away from empiricism little by little; like all other sciences, it will get away by the scientific method. This deep conviction sustains and guides my scientific life. I am deaf to the physicians who ask us to explain measles and scarlet fever experimentally, and who believe they can find in them an argument against using the experimental method. These discouraging, negative objections generally come from systematic or lazy minds that prefer resting on their systems or sleeping in the dark, to working and making an effort to get away. The different branches of physico-chemical science were elucidated only gradually, step by step, by the experimental method, and they still have to-day obscure parts which we are studying with the help of the same method. In spite of all the obstacles that it meets, medicine will follow the same course; it will follow it necessarily. In extolling the introduction of the experimental method into medicine, I am therefore only trying to guide men's minds toward a goal that science is instinctively and unconsciously pursuing,—a goal that it will more quickly and certainly reach if it can succeed in seeing it clearly. Time will then do the rest. Of course, we shall not see scientific medicine blossoming in our day, but that is man's lot; those who sow and laboriously cultivate the field of science are not also destined to reap the harvest.

To sum up, experimental medicine, as we conceive it, includes the problem of medicine as a whole and comprises both the theory and the practice of medicine. But when I said that every physician should be an experimenter, I did not mean to suggest that each one should cultivate the whole extent of experimental medicine. Of necessity, there will always be physicians especially devoting themselves to physiological experiments, others to investigation of normal and pathological anatomy, others to surgical or medical practice, etc. This splitting up is not bad for the progress of science; on the contrary, practical specialties are an excellent thing for science, properly speaking, but on the condition that men devoting themselves to the investigation of a special part of medicine be so educated as to be conversant with experimental medicine as a whole, and to know the place which the special science they cultivate should occupy in that whole. By specializing in this way, they will direct their studies so as to contribute to the progress of scientific or experimental medicine. Practical studies and theoretic studies will thus work toward the same object; that is all that we can ask in a science, like medicine, which is forced to be ceaselessly in action before it is fully established.

Experimental or scientific medicine is tending on every side to establish itself on the basis of physiology. The tendency of studies published every day, whether in France or abroad, furnishes unmistakable proof. In my research and teaching at the Collège de France, I unfold every idea that can help or encourage this tendency in medicine. I consider this my duty as a man of science and professor of medicine at the Collège de France. In fact, the Collège de France is by no means a medical faculty in which every part of medicine should be treated systematically. By the very nature of its establishment, the Collège de France should always be in the forefront of science, embodying its movement and its tendencies. Consequently the course in medicine with which I am entrusted must embody the part of medical science which is by way of the greatest present development, and which involves the rest in its evolution. I have already explained myself at length on the proper character of the course in medicine at the Collège de France; I shall not return to that.[3] Let me simply say that, while I acknowledge that the experimental trend of science must be slow to establish itself because of difficulties inherent in the complexity of medicine, we must recognize that it is now a definite trend. In fact, this has not been brought about by the ephemeral influence of some personal system or other; it results from the scientific evolution of medicine itself. My convictions, in this respect, are what I am seeking to impress on the

---

[3] Claude Bernard, *Leçons de physiologie expérimentale appliquée à la médecine* (*faites au Collège de France*). First lesson, Paris, 1857. *Cours de médecine au Collège de France.* First lesson, Paris, 1855.

minds of the young physicians attending my courses at the Collège de France. I try to show them that they are all called to contribute their share to the increase and development of experimental or scientific medicine. For that reason I ask them to familiarize themselves with the modern methods of investigation put in use in anatomical, physiological, pathological and therapeutic science; because these various branches of medicine must always remain inseparably united in theory and in practice. I tell those whose path leads them toward theory or toward pure science, never to lose sight of the medical problem, which is to preserve health and cure disease. I tell those whose career, on the contrary, guides them toward practice, never to forget that if theory is meant to enlighten practice, practice in turn should be of use to science. Physicians thoroughly imbued with these ideas will always keep their interest in the progress of science, at the same time that they do their duty as practitioners. In noting accurately and acutely the interesting cases that present themselves, they will understand how fully science may profit by them. Experimental scientific medicine will thus become the achievement of us all; and every one of us, even if he be only a simple country doctor, will make his own useful contribution.

Returning now to the title of this long section, I conclude that empirical medicine and experimental medicine are far from being incompatible, but on the contrary must be intimately united; for both are indispensable in building up experimental medicine. I think that this conclusion is well established by all that has gone before.

IV. Experimental Medicine Does Not Correspond to Any Medical Doctrine or Any Philosophic System

We said that experimental medicine is not a new system of medicine, but on the contrary is the negation of all systems. In fact, the advent of experimental medicine will cause all individual views to disappear from the science, to be replaced by impersonal and general theories which, as in other sciences, will be only a regular and logical coördination of facts furnished by experience.

Scientific medicine is certainly not yet well established to-day; but thanks to the experimental method which is permeating it more and more, it is tending to become an exact science. Medicine is in transition; the day of personal doctrines and systems is past, and little by little they will be replaced by theories embodying the present state of the science and showing from that point of view the results of all our efforts. But that must not make us believe that theories are ever absolute truths; they may always be improved, and so are always mobile. That is why I have been careful to say that we must not, as men often do, confuse advancing and perfectible progressive theories, which may be improved, with scientific methods and principles that are fixed and unshakable. We must remember that the one unchangeable scientific principle, in medicine as well as in the other experimental sciences, is the absolute determinism of phenomena. We gave the name of determinism to the immediate or determining cause of phenomena. We never act on the essence of natural phenomena, but only on their determining causes; and because we act thus, determinism differs from fatalism, on which we cannot act. Fatalism assumes that the manifestation of any phenomenon is necessary and independent of its conditions, while determinism is the condition necessary to a phenomenon, whose manifestation is free. When search for the causes determining phenomena is once posited as the fundamental principle of the experimental method, materialism, spiritualism, inert matter and living matter cease to exist; only phenomena are left, whose conditions we must determine, i.e., the conditions which play the part of immediate cause. Scientific determinism ceases here; there are only words beyond, which are of course necessary, but which may delude us if we are not constantly on guard against the traps which our minds perpetually set for themselves.

As experimental medicine, like all the experimental sciences, should not go beyond phenomena, it does not need to be tied to any system; it is neither vitalistic, nor animistic, nor organistic, nor solidistic, nor humoral; it is simply the science which tries to reach the immediate causes of vital phenomena in the healthy and in the morbid state. It has no reason, in fact, to encumber itself with systems, none of which can ever embody the truth.

In this connection it may be useful to recall, in a few words, the essential characteristics of the scientific method and to show how the ideas derived from it differ from systematic and doctrinal ideas. In the experimental method we never make experiments except to see or to prove, i.e., to control or verify. As a scientific method, the experimental method rests wholly on the experimental verification of a scientific hypothesis. We obtain this verification with the help, sometimes of a fresh observation (observational science), sometimes of an experiment (experimental science). In the experimental method, the hypothesis is a scientific idea that we submit to experiment. Scientific invention consists in the creation of fortunate and fertile hypotheses; these are suggested by the feeling or even the genius of the men of science who create them.

When an hypothesis is submitted to the experimental method, it becomes a theory, while if it is submitted to logic alone, it becomes a system. A system, then, is an hypothesis with which we have connected the facts logically with the help of reason, but without experimental, critical verification. A theory is a verified hypothesis, after it has been submitted to the control of reason and experimental criticism. The soundest theory is one that has been verified by the greatest number of facts. But to remain valid, a theory must be continually altered to keep pace with the progress of science and must be constantly resubmitted to verification and criticism as new facts appear.

If we consider a theory perfect and stop verifying it by daily scientific experience, it becomes a doctrine. A doctrine, then, is a theory which we regard as immutable, which we take as a starting point for later deduction, and which we believe we are no longer obliged to submit to experimental verification.

In a word, systems and doctrines in medicine are hypothetical or theoretic ideas transformed into immutable principles. This sort of method belongs essentially to scholasticism and differs radically from the experimental method. These two methods of the mind, indeed, are contradictory. Systems and doctrines proceed by affirmation and purely logical deduction; the experimental method always proceeds by doubt and experimental verification. Systems and doctrines are individual; they are meant to be immutable and to preserve their personal aspect. The experimental method, on the other hand, is impersonal; it destroys individuality by uniting and sacrificing everyone's particular ideas, and turning them to the advantage of universal truth as established with the help of the experimental criterion. It advances slowly and laboriously and in this respect will always be less pleasing to the mind. Systems, on the contrary, are alluring because they give us a science absolutely regulated by logic alone; and that frees us from studying and makes medicine easy. Experimental medicine, then, is anti-systematic and anti-doctrinal by nature, or rather it is free and independent in its essence and does not try to attach itself to any kind of medical system.

What I have just been saying about medical systems, I can apply to philosophic systems. Experimental medicine (like all experimental sciences, for that matter) does not need to be attached to any philosophic system. A physiologist's rôle, like every scientific man's, is to seek truth for its own sake, without wishing to use it to control one system of philosophy or another. When a man of science takes a philosophic system as his base in pursuing a scientific investigation, he goes astray in regions that are too far from reality, or else the system gives his mind a sort of false confidence and an inflexibility out of harmony with the freedom and suppleness that experimenters should always maintain in their researches. We must therefore carefully avoid every species of system, because systems are not found in nature, but only in the mind of man. Positivism, like the philosophic systems which it rejects in the name of science, has the fault of being a system. Now, to find truth, men of science need only stand face to face with nature, and in following experimental medicine, question her with the help of more and more perfect means of investigation. In this case, I think that the best philosophic system consists in not having any.

As an experimenter, then, I avoid philosophic systems; but I cannot for that reason reject the philosophic spirit which, without being anywhere, is everywhere and, without belonging to any system, ought to reign, not only over all science but over all human knowledge. So even while avoiding philosophic systems, I like philosophers and greatly enjoy their converse. Indeed, from the scientific point of view, philosophy embodies the eternal aspiration of human reason toward knowledge of the unknown. Therefore philosophers always live in controversial questions and in lofty regions, the upper boundaries of science. Hence they impart to scientific thought an enlivening and ennobling motion; they develop and fortify the mind by general intellectual exercise, while ceaselessly bearing it toward the inexhaustible solution of great problems; thus they nourish a kind of thirst for the unknown; the sacred fire of research must therefore never be extinguished in men of science.

Ardent desire for knowledge, in fact, is the one motive attracting and supporting investigators in their efforts; and just this knowledge, really grasped and yet always flying before them, becomes at once their sole torment and sole happiness. Those who do not know the torment of the unknown cannot have the joy of discovery which is certainly the liveliest that the mind of man can ever feel. But by a whim of our nature, the joy of discovery, so sought and hoped for, vanishes as soon as found. It is but a flash whose gleam discovers for us fresh horizons, toward which our insatiate curiosity repairs with still more ardor. Thus, even in science itself, the known loses its attraction, while the unknown is always full of charm. Therefore the minds that rise and become really great are never self-satisfied, but still continue to strive. The feeling, about which I am speaking now, is familiar to men of science and to philosophers. This is the feeling that made Priestley [4] say that each discovery we make shows us many others that should be made; this is the feeling which Pascal expressed in somewhat paradoxical form, when he said: "We are in search never of things, but of the search for things." Yet truth itself is surely what concerns us and, if we are still in search of it, that is because the part which we have so far found cannot satisfy us. In our investigations, we should else be performing the useless and endless labor pictured in the fable of Sisyphus, ever rolling up the rock which continually falls back to its starting point. This comparison is not scientifically correct: a man of science rises ever, in seeking truth; and if he never finds it in its wholeness, he discovers nevertheless very significant fragments; and these fragments of universal truth are precisely what constitute science.

Men of science, then, do not seek for the pleasure of seeking; they seek the truth to possess it, and they possess it already within the limits expressed in the present state of the sciences. But men of science must not halt on the road; they must climb ever higher and strive toward perfection; they must always seek, as long as they see anything to be found. Without constant stimulation by the spur of the unknown, without constantly recurring thirst, it might be feared that men of science would become system-ridden in their acquirements and their knowledge. Then science would halt through intellectual inertness, just as minerals, in saturated solution,

[4] Priestley, *Experiments and Observations on Different Kinds of Air.* Introduction, p. 15.

become chemically inert and crystallize. We must therefore prevent our minds from becoming too much absorbed in the known parts of any particular science or dragging themselves along the ground and losing sight of questions still to be solved. By ceaselessly stirring the inexhaustible mass of unsolved questions, philosophy stimulates and maintains this healthful movement in science. For only the indeterminate belongs to philosophy, in the restricted sense in which I am here considering it, while the determinate necessarily falls into the realm of science. I can no more accept a philosophy, then, which tries to assign boundaries to science, than a science which claims to suppress philosophic truths that are at present outside its own domain. True science suppresses nothing, but goes on searching, and is undisturbed in looking straight at things that it does not yet understand. If we denied these facts, we should not suppress them; we should only be shutting our eyes and believing there was no light; we should be sharing the delusion of the ostrich which believes it banishes danger by hiding its head in the sand. In my opinion the true scientific spirit is that whose high aspirations fertilize the sciences and draw them on in search of truths which are still beyond them, but which must not be suppressed, because they have been attacked by stronger and more delicate philosophic minds. Has this aspiration of the human spirit any end,—will it find its limit? That, I cannot know; but meantime, as I said above, men of science can do no better than to push steadily forward, because they can always go forward.

One of the greatest obstacles to the free and universal movement of human knowledge is the tendency that leads different kinds of knowledge to separate into systems. This is not a consequence of things in themselves, because everything in nature is connected with everything else and nothing should be viewed in the isolation of a system; but the feeble yet dominating tendency of our minds leads us to absorb other kinds of knowledge into our personal systems. A science that halted in a system would remain stationary and would be isolated, because systematization is really a scientific encysting, and every encysted part of an organism ceases to take part in the organism's general life. Systems therefore strive to enslave the human mind, and, in my opinion, their only ascertainable use is to promote conflicts which destroy them, by stirring and stimulating the vitality of science. We must try, indeed, to break the fetters of

philosophic and scientific systems, as we would break the chains of
intellectual slavery. Truth, if we can find it, belongs to every
system; and to discover it, experimenters need free movement on
every side, without feeling themselves stopped by the barriers of any
system. Philosophy and science, then, must never be systematic:
without trying to dominate one another, they must unite. Their
separation could only be harmful to the progress of human knowledge.
Striving ever upward, philosophy makes science rise toward the cause
or the source of things. It shows science that there are questions
beyond it, torturing humanity, which it has not yet solved. Solid
union between science and philosophy is useful to both: it lifts the
one and confines the other. But if the bonds uniting philosophy to
science should break, philosophy, lacking the support or the counter-
poise of science would rise out of sight and be lost in the clouds, while
science, without guidance and without high aspiration, would sail
at random.

But if philosophy, instead of contenting itself with this fraternal
union, tried to enter the household of science and dogmatically lord
its productions and its methods of manifestation, then their understanding would cease. Claiming to absorb the special discoveries
of a science into any philosophic system would, in fact, be a delusion.
For making scientific observations, experiments and discoveries, philosophic method and procedure are vague and powerless; the only
means available for that are scientific methods and procedures that
can be known only by experimenters, men of science or philosophers,
practising some definite science. The different kinds of human
knowledge are so entangled and so interdependent in their evolution,
that we cannot possibly believe that any individual influence can
advance them unless the elements of progress are present in the
scientific soil itself. While recognizing the superiority of great
men, I therefore still think that, even in their special or general influence on science, they are always necessarily more or less a function
of their time. It is the same with philosophers: they can only follow
the movement of the human mind, and they contribute to its advance,
only by opening the path of progress wider. But in that, they are
an expression of their time. No philosopher, coming at a moment
when science takes a fertile turn, should create a system, then, in
harmony with the movement of science, and afterward cry out that
all the scientific progress of his day is due to the influence of his

system. In a word, if men of science are useful to philosophers, and philosophers to men of science, men of science remain free, none the less, and masters in their own house; as for myself, I think that men of science achieve their discoveries, their theories and their science apart from philosophers. If we meet with incredulity with regard to this, we can perhaps easily prove that, as J. de Maistre says, those who make the most discoveries in science know Bacon [5] least, while those who read and ponder him, like Bacon himself, have poor success. For scientific methods and processes are learned, in fact, only in laboratories, where experimenters grapple with the problems of nature; the young must be guided thither first of all; men of riper age have as their portion erudition and scientific criticism which can bear fruit only when we have begun our initiation into science in its true sanctuary, the laboratory. Processes of reasoning should endlessly vary for experimenters, according to the different sciences and to the more or less difficult questions to which they apply them. Only scientific men, and indeed scientific men specializing in each science, can take up such questions, because a naturalist's mind is not a physiologist's mind, any more than a chemist's mind is a physicist's. As for Bacon and the other more modern philosophers who try a general systematization of precepts for scientific research, they may seem alluring to people who look at science only from a distance; but works like theirs are of no use to experienced scientists; and by false simplification of things, they mislead men who wish to devote themselves to cultivating science. What is more, they embarrass them by burdening the mind with vague and inapplicable precepts that we must hasten to forget if we wish to become true experimenters.

I have said that scientific men and experimenters can be educated only in special laboratories of the sciences they wish to cultivate and that precepts are useful only when derived from the details of experimental practice in some definite science. In this Introduction, I have tried to give as exact an idea as possible of physiological science and of experimental medicine. However, I am far fom presuming to believe that I have given rules and precepts which experimenters should follow rigorously and absolutely. I have tried merely to study the nature of the problems to be solved in the experimental science of living beings, so that everyone might thoroughly under-

[5] J. de Maistre. *Examen de la philosophie de Bacon*, Vol. I, p. 81.

stand the scientific questions belonging to the domain of biology and know the means which that science now has to attack them. I have quoted examples of investigation, but have been very careful not to give needless explanations or to formulate a single or absolute rule; because I think a teacher's rôle should be limited to clearly showing his pupil the goal that a science sets itself and to pointing out all possible means at his disposal for reaching it. But a teacher should then leave his pupil free to move about in his own way and, according to his own nature, to reach his goal, only coming to his aid if he sees that he is going astray. I believe, in a word, that the true scientific method confines the mind without suffocating it, leaves it as far as possible face to face with itself, and guides it, while respecting the creative originality and the spontaneity which are its most precious qualities. Science goes forward only through new ideas and through creative or original power of thought. In education we must, therefore, take care that knowledge which should arm the mind does not overwhelm it by its weight, and that rules, intended to support weak parts of the mind, do not atrophy the strong and fertile parts. I need not enter into further explanations here; I have had to limit myself by forewarning biological science and experimental medicine against exaggerating the importance of erudition and against invasion and domination by systems; because sciences submitting to these would lose their fertility and would abandon the independence and freedom of mind essential to the progress of humanity.

A CATALOG OF
SELECTED DOVER BOOKS
IN ALL FIELDS OF INTEREST

A CATALOG OF SELECTED DOVER BOOKS IN ALL FIELDS OF INTEREST

CONCERNING THE SPIRITUAL IN ART, Wassily Kandinsky. Pioneering work by father of abstract art. Thoughts on color theory, nature of art. Analysis of earlier masters. 12 illustrations. 80pp. of text. 5⅜ × 8½. 23411-8 Pa. $2.50

LEONARDO ON THE HUMAN BODY, Leonardo da Vinci. More than 1200 of Leonardo's anatomical drawings on 215 plates. Leonardo's text, which accompanies the drawings, has been translated into English. 506pp. 8⅜ × 11¾. 24483-0 Pa. $10.95

GOBLIN MARKET, Christina Rossetti. Best-known work by poet comparable to Emily Dickinson, Alfred Tennyson. With 46 delightfully grotesque illustrations by Laurence Housman. 64pp. 4 × 6¾. 24516-0 Pa. $2.50

THE HEART OF THOREAU'S JOURNALS, edited by Odell Shepard. Selections from *Journal,* ranging over full gamut of interests. 228pp. 5⅜ × 8½. 20741-2 Pa. $4.50

MR. LINCOLN'S CAMERA MAN: MATHEW B. BRADY, Roy Meredith. Over 300 Brady photos reproduced directly from original negatives, photos. Lively commentary. 368pp. 8⅜ × 11¼. 23021-X Pa. $11.95

PHOTOGRAPHIC VIEWS OF SHERMAN'S CAMPAIGN, George N. Barnard. Reprint of landmark 1866 volume with 61 plates: battlefield of New Hope Church, the Etawah Bridge, the capture of Atlanta, etc. 80pp. 9 × 12. 23445-2 Pa. $6.00

A SHORT HISTORY OF ANATOMY AND PHYSIOLOGY FROM THE GREEKS TO HARVEY, Dr. Charles Singer. Thoroughly engrossing non-technical survey. 270 illustrations. 211pp. 5⅜ × 8½. 20389-1 Pa. $4.50

REDOUTE ROSES IRON-ON TRANSFER PATTERNS, Barbara Christopher. Redouté was botanical painter to the Empress Josephine; transfer his famous roses onto fabric with these 24 transfer patterns. 80pp. 8¼ × 10⅞. 24292-7 Pa. $3.50

THE FIVE BOOKS OF ARCHITECTURE, Sebastiano Serlio. Architectural milestone, first (1611) English translation of Renaissance classic. Unabridged reproduction of original edition includes over 300 woodcut illustrations. 416pp. 9⅜ × 12¼. 24349-4 Pa. $14.95

CARLSON'S GUIDE TO LANDSCAPE PAINTING, John F. Carlson. Authoritative, comprehensive guide covers, every aspect of landscape painting. 34 reproductions of paintings by author; 58 explanatory diagrams. 144pp. 8⅜ × 11. 22927-0 Pa. $4.95

101 PUZZLES IN THOUGHT AND LOGIC, C.R. Wylie, Jr. Solve murders, robberies, see which fishermen are liars—purely by reasoning! 107pp. 5⅜ × 8½. 20367-0 Pa. $2.00

TEST YOUR LOGIC, George J. Summers. 50 more truly new puzzles with new turns of thought, new subtleties of inference. 100pp. 5⅜ × 8½. 22877-0 Pa. $2.25

THE MURDER BOOK OF J.G. REEDER, Edgar Wallace. Eight suspenseful stories by bestselling mystery writer of 20s and 30s. Features the donnish Mr. J.G. Reeder of Public Prosecutor's Office. 128pp. 5⅜ × 8½. (Available in U.S. only)
24374-5 Pa. $3.50

ANNE ORR'S CHARTED DESIGNS, Anne Orr. Best designs by premier needlework designer, all on charts: flowers, borders, birds, children, alphabets, etc. Over 100 charts, 10 in color. Total of 40pp. 8¼ × 11.
23704-4 Pa. $2.25

BASIC CONSTRUCTION TECHNIQUES FOR HOUSES AND SMALL BUILDINGS SIMPLY EXPLAINED, U.S. Bureau of Naval Personnel. Grading, masonry, woodworking, floor and wall framing, roof framing, plastering, tile setting, much more. Over 675 illustrations. 568pp. 6½ × 9¼.
20242-9 Pa. $8.95

MATISSE LINE DRAWINGS AND PRINTS, Henri Matisse. Representative collection of female nudes, faces, still lifes, experimental works, etc., from 1898 to 1948. 50 illustrations. 48pp. 8⅜ × 11¼.
23877-6 Pa. $2.50

HOW TO PLAY THE CHESS OPENINGS, Eugene Znosko-Borovsky. Clear, profound examinations of just what each opening is intended to do and how opponent can counter. Many sample games. 147pp. 5⅜ × 8½.
22795-2 Pa. $2.95

DUPLICATE BRIDGE, Alfred Sheinwold. Clear, thorough, easily followed account: rules, etiquette, scoring, strategy, bidding; Goren's point-count system, Blackwood and Gerber conventions, etc. 158pp. 5⅜ × 8½.
22741-3 Pa. $3.00

SARGENT PORTRAIT DRAWINGS, J.S. Sargent. Collection of 42 portraits reveals technical skill and intuitive eye of noted American portrait painter, John Singer Sargent. 48pp. 8¼ × 11⅛.
24524-1 Pa. $2.95

ENTERTAINING SCIENCE EXPERIMENTS WITH EVERYDAY OBJECTS, Martin Gardner. Over 100 experiments for youngsters. Will amuse, astonish, teach, and entertain. Over 100 illustrations. 127pp. 5⅜ × 8½.
24201-3 Pa. $2.50

TEDDY BEAR PAPER DOLLS IN FULL COLOR: A Family of Four Bears and Their Costumes, Crystal Collins. A family of four Teddy Bear paper dolls and nearly 60 cut-out costumes. Full color, printed one side only. 32pp. 9¼ × 12¼.
24550-0 Pa. $3.50

NEW CALLIGRAPHIC ORNAMENTS AND FLOURISHES, Arthur Baker. Unusual, multi-useable material: arrows, pointing hands, brackets and frames, ovals, swirls, birds, etc. Nearly 700 illustrations. 80pp. 8⅜ × 11¼.
24095-9 Pa. $3.75

DINOSAUR DIORAMAS TO CUT & ASSEMBLE, M. Kalmenoff. Two complete three-dimensional scenes in full color, with 31 cut-out animals and plants. Excellent educational toy for youngsters. Instructions; 2 assembly diagrams. 32pp. 9¼ × 12¼.
24541-1 Pa. $3.95

SILHOUETTES: A PICTORIAL ARCHIVE OF VARIED ILLUSTRATIONS, edited by Carol Belanger Grafton. Over 600 silhouettes from the 18th to 20th centuries. Profiles and full figures of men, women, children, birds, animals, groups and scenes, nature, ships, an alphabet. 144pp. 8⅜ × 11¼.
23781-8 Pa. $4.95

25 KITES THAT FLY, Leslie Hunt. Full, easy-to-follow instructions for kites made from inexpensive materials. Many novelties. 70 illustrations. 110pp. 5⅜ × 8½.
22550-X Pa. $2.25

PIANO TUNING, J. Cree Fischer. Clearest, best book for beginner, amateur. Simple repairs, raising dropped notes, tuning by easy method of flattened fifths. No previous skills needed. 4 illustrations. 201pp. 5⅜ × 8½. 23267-0 Pa. $3.50

EARLY AMERICAN IRON-ON TRANSFER PATTERNS, edited by Rita Weiss. 75 designs, borders, alphabets, from traditional American sources. 48pp. 8¼ × 11.
23162-3 Pa. $1.95

CROCHETING EDGINGS, edited by Rita Weiss. Over 100 of the best designs for these lovely trims for a host of household items. Complete instructions, illustrations. 48pp. 8¼ × 11. 24031-2 Pa. $2.25

FINGER PLAYS FOR NURSERY AND KINDERGARTEN, Emilie Poulsson. 18 finger plays with music (voice and piano); entertaining, instructive. Counting, nature lore, etc. Victorian classic. 53 illustrations. 80pp. 6½ × 9¼. 22588-7 Pa. $1.95

BOSTON THEN AND NOW, Peter Vanderwarker. Here in 59 side-by-side views are photographic documentations of the city's past and present. 119 photographs. Full captions. 122pp. 8¼ × 11. 24312-5 Pa. $6.95

CROCHETING BEDSPREADS, edited by Rita Weiss. 22 patterns, originally published in three instruction books 1939-41. 39 photos, 8 charts. Instructions. 48pp. 8¼ × 11. 23610-2 Pa. $2.00

HAWTHORNE ON PAINTING, Charles W. Hawthorne. Collected from notes taken by students at famous Cape Cod School; hundreds of direct, personal *apercus*, ideas, suggestions. 91pp. 5⅜ × 8½. 20653-X Pa. $2.50

THERMODYNAMICS, Enrico Fermi. A classic of modern science. Clear, organized treatment of systems, first and second laws, entropy, thermodynamic potentials, etc. Calculus required. 160pp. 5⅜ × 8½. 60361-X Pa. $4.00

TEN BOOKS ON ARCHITECTURE, Vitruvius. The most important book ever written on architecture. Early Roman aesthetics, technology, classical orders, site selection, all other aspects. Morgan translation. 331pp. 5⅜ × 8½. 20645-9 Pa. $5.50

THE CORNELL BREAD BOOK, Clive M. McCay and Jeanette B. McCay. Famed high-protein recipe incorporated into breads, rolls, buns, coffee cakes, pizza, pie crusts, more. Nearly 50 illustrations. 48pp. 8¼ × 11. 23995-0 Pa. $2.00

THE CRAFTSMAN'S HANDBOOK, Cennino Cennini. 15th-century handbook, school of Giotto, explains applying gold, silver leaf; gesso; fresco painting, grinding pigments, etc. 142pp. 6⅛ × 9¼. 20054-X Pa. $3.50

FRANK LLOYD WRIGHT'S FALLINGWATER, Donald Hoffmann. Full story of Wright's masterwork at Bear Run, Pa. 100 photographs of site, construction, and details of completed structure. 112pp. 9¼ × 10. 23671-4 Pa. $6.50

OVAL STAINED GLASS PATTERN BOOK, C. Eaton. 60 new designs framed in shape of an oval. Greater complexity, challenge with sinuous cats, birds, mandalas framed in antique shape. 64pp. 8¼ × 11. 24519-5 Pa. $3.50

THE BOOK OF WOOD CARVING, Charles Marshall Sayers. Still finest book for beginning student. Fundamentals, technique; gives 34 designs, over 34 projects for panels, bookends, mirrors, etc. 33 photos. 118pp. 7¾ × 10⅝. 23654-4 Pa. $3.95

CARVING COUNTRY CHARACTERS, Bill Higginbotham. Expert advice for beginning, advanced carvers on materials, techniques for creating 18 projects— mirthful panorama of American characters. 105 illustrations. 80pp. 8⅜ × 11.
24135-1 Pa. $2.50

300 ART NOUVEAU DESIGNS AND MOTIFS IN FULL COLOR, C.B. Grafton. 44 full-page plates display swirling lines and muted colors typical of Art Nouveau. Borders, frames, panels, cartouches, dingbats, etc. 48pp. 9⅜ × 12¼.
24354-0 Pa. $6.00

SELF-WORKING CARD TRICKS, Karl Fulves. Editor of *Pallbearer* offers 72 tricks that work automatically through nature of card deck. No sleight of hand needed. Often spectacular. 42 illustrations. 113pp. 5⅜ × 8½. 23334-0 Pa. $3.50

CUT AND ASSEMBLE A WESTERN FRONTIER TOWN, Edmund V. Gillon, Jr. Ten authentic full-color buildings on heavy cardboard stock in H-O scale. Sheriff's Office and Jail, Saloon, Wells Fargo, Opera House, others. 48pp. 9¼ × 12¼.
23736-2 Pa. $3.95

CUT AND ASSEMBLE AN EARLY NEW ENGLAND VILLAGE, Edmund V. Gillon, Jr. Printed in full color on heavy cardboard stock. 12 authentic buildings in H-O scale: Adams home in Quincy, Mass., Oliver Wight house in Sturbridge, smithy, store, church, others. 48pp. 9¼ × 12¼. 23536-X Pa. $3.95

THE TALE OF TWO BAD MICE, Beatrix Potter. Tom Thumb and Hunca Munca squeeze out of their hole and go exploring. 27 full-color Potter illustrations. 59pp. 4¼ × 5½. (Available in U.S. only) 23065-1 Pa. $1.50

CARVING FIGURE CARICATURES IN THE OZARK STYLE, Harold L. Enlow. Instructions and illustrations for ten delightful projects, plus general carving instructions. 22 drawings and 47 photographs altogether. 39pp. 8⅜ × 11.
23151-8 Pa. $2.50

A TREASURY OF FLOWER DESIGNS FOR ARTISTS, EMBROIDERERS AND CRAFTSMEN, Susan Gaber. 100 garden favorites lushly rendered by artist for artists, craftsmen, needleworkers. Many form frames, borders. 80pp. 8¼ × 11.
24096-7 Pa. $3.50

CUT & ASSEMBLE A TOY THEATER/THE NUTCRACKER BALLET, Tom Tierney. Model of a complete, full-color production of Tchaikovsky's classic. 6 backdrops, dozens of characters, familiar dance sequences. 32pp. 9⅜ × 12¼.
24194-7 Pa. $4.50

ANIMALS: 1,419 COPYRIGHT-FREE ILLUSTRATIONS OF MAMMALS, BIRDS, FISH, INSECTS, ETC., edited by Jim Harter. Clear wood engravings present, in extremely lifelike poses, over 1,000 species of animals. 284pp. 9 × 12.
23766-4 Pa. $9.95

MORE HAND SHADOWS, Henry Bursill. For those at their 'finger ends," 16 more effects—Shakespeare, a hare, a squirrel, Mr. Punch, and twelve more—each explained by a full-page illustration. Considerable period charm. 30pp. 6½ × 9¼.
21384-6 Pa. $1.95

SURREAL STICKERS AND UNREAL STAMPS, William Rowe. 224 haunting, hilarious stamps on gummed, perforated stock, with images of elephants, geisha girls, George Washington, etc. 16pp. one side. 8¼ × 11.   24371-0 Pa. $3.50

GOURMET KITCHEN LABELS, Ed Sibbett, Jr. 112 full-color labels (4 copies each of 28 designs). Fruit, bread, other culinary motifs. Gummed and perforated. 16pp. 8¼ × 11.   24087-8 Pa. $2.95

PATTERNS AND INSTRUCTIONS FOR CARVING AUTHENTIC BIRDS, H.D. Green. Detailed instructions, 27 diagrams, 85 photographs for carving 15 species of birds so life-like, they'll seem ready to fly! 8¼ × 11.   24222-6 Pa. $2.75

FLATLAND, E.A. Abbott. Science-fiction classic explores life of 2-D being in 3-D world. 16 illustrations. 103pp. 5⅜ × 8.   20001-9 Pa. $2.00

DRIED FLOWERS, Sarah Whitlock and Martha Rankin. Concise, clear, practical guide to dehydration, glycerinizing, pressing plant material, and more. Covers use of silica gel. 12 drawings. 32pp. 5⅜ × 8½.   21802-3 Pa. $1.00

EASY-TO-MAKE CANDLES, Gary V. Guy. Learn how easy it is to make all kinds of decorative candles. Step-by-step instructions. 82 illustrations. 48pp. 8¼ × 11.
23881-4 Pa. $2.50

SUPER STICKERS FOR KIDS, Carolyn Bracken. 128 gummed and perforated full-color stickers: GIRL WANTED, KEEP OUT, BORED OF EDUCATION, X-RATED, COMBAT ZONE, many others. 16pp. 8¼ × 11.   24092-4 Pa. $2.50

CUT AND COLOR PAPER MASKS, Michael Grater. Clowns, animals, funny faces...simply color them in, cut them out, and put them together, and you have 9 paper masks to play with and enjoy. 32pp. 8¼ × 11.   23171-2 Pa. $2.25

A CHRISTMAS CAROL: THE ORIGINAL MANUSCRIPT, Charles Dickens. Clear facsimile of Dickens manuscript, on facing pages with final printed text. 8 illustrations by John Leech, 4 in color on covers. 144pp. 8⅜ × 11¼.
20980-6 Pa. $5.95

CARVING SHOREBIRDS, Harry V. Shourds & Anthony Hillman. 16 full-size patterns (all double-page spreads) for 19 North American shorebirds with step-by-step instructions. 72pp. 9¼ × 12¼.   24287-0 Pa. $4.95

THE GENTLE ART OF MATHEMATICS, Dan Pedoe. Mathematical games, probability, the question of infinity, topology, how the laws of algebra work, problems of irrational numbers, and more. 42 figures. 143pp. 5⅜ × 8½. (EBE)
22949-1 Pa. $3.50

READY-TO-USE DOLLHOUSE WALLPAPER, Katzenbach & Warren, Inc. Stripe, 2 floral stripes, 2 allover florals, polka dot; all in full color. 4 sheets (350 sq. in.) of each, enough for average room. 48pp. 8¼ × 11.   23495-9 Pa. $2.95

MINIATURE IRON-ON TRANSFER PATTERNS FOR DOLLHOUSES, DOLLS, AND SMALL PROJECTS, Rita Weiss and Frank Fontana. Over 100 miniature patterns: rugs, bedspreads, quilts, chair seats, etc. In standard dollhouse size. 48pp. 8¼ × 11.   23741-9 Pa. $1.95

THE DINOSAUR COLORING BOOK, Anthony Rao. 45 renderings of dinosaurs, fossil birds, turtles, other creatures of Mesozoic Era. Scientifically accurate. Captions. 48pp. 8¼ × 11.   24022-3 Pa. $2.25

JAPANESE DESIGN MOTIFS, Matsuya Co. Mon, or heraldic designs. Over 4000 typical, beautiful designs: birds, animals, flowers, swords, fans, geometrics; all beautifully stylized. 213pp. 11⅜ × 8¼. 22874-6 Pa. $7.95

THE TALE OF BENJAMIN BUNNY, Beatrix Potter. Peter Rabbit's cousin coaxes him back into Mr. McGregor's garden for a whole new set of adventures. All 27 full-color illustrations. 59pp. 4¼ × 5½. (Available in U.S. only) 21102-9 Pa. $1.50

THE TALE OF PETER RABBIT AND OTHER FAVORITE STORIES BOXED SET, Beatrix Potter. Seven of Beatrix Potter's best-loved tales including Peter Rabbit in a specially designed, durable boxed set. 4¼ × 5½. Total of 447pp. 158 color illustrations. (Available in U.S. only) 23903-9 Pa. $10.80

PRACTICAL MENTAL MAGIC, Theodore Annemann. Nearly 200 astonishing feats of mental magic revealed in step-by-step detail. Complete advice on staging, patter, etc. Illustrated. 320pp. 5⅜ × 8½. 24426-1 Pa. $5.95

CELEBRATED CASES OF JUDGE DEE (DEE GOONG AN), translated by Robert Van Gulik. Authentic 18th-century Chinese detective novel; Dee and associates solve three interlocked cases. Led to van Gulik's own stories with same characters. Extensive introduction. 9 illustrations. 237pp. 5⅜ × 8½. 23337-5 Pa. $4.50

CUT & FOLD EXTRATERRESTRIAL INVADERS THAT FLY, M. Grater. Stage your own lilliputian space battles.By following the step-by-step instructions and explanatory diagrams you can launch 22 full-color fliers into space. 36pp. 8¼ × 11. 24478-4 Pa. $2.95

CUT & ASSEMBLE VICTORIAN HOUSES, Edmund V. Gillon, Jr. Printed in full color on heavy cardboard stock, 4 authentic Victorian houses in H-O scale: Italian-style Villa, Octagon, Second Empire, Stick Style. 48pp. 9¼ × 12¼. 23849-0 Pa. $3.95

BEST SCIENCE FICTION STORIES OF H.G. WELLS, H.G. Wells. Full novel *The Invisible Man,* plus 17 short stories: "The Crystal Egg," "Aepyornis Island," "The Strange Orchid," etc. 303pp. 5⅜ × 8½. (Available in U.S. only) 21531-8 Pa. $4.95

TRADEMARK DESIGNS OF THE WORLD, Yusaku Kamekura. A lavish collection of nearly 700 trademarks, the work of Wright, Loewy, Klee, Binder, hundreds of others. 160pp. 8¾ × 8. (Available in U.S. only) 24191-2 Pa. $5.00

THE ARTIST'S AND CRAFTSMAN'S GUIDE TO REDUCING, ENLARGING AND TRANSFERRING DESIGNS, Rita Weiss. Discover, reduce, enlarge, transfer designs from any objects to any craft project. 12pp. plus 16 sheets special graph paper. 8¼ × 11. 24142-4 Pa. $3.25

TREASURY OF JAPANESE DESIGNS AND MOTIFS FOR ARTISTS AND CRAFTSMEN, edited by Carol Belanger Grafton. Indispensable collection of 360 traditional Japanese designs and motifs redrawn in clean, crisp black-and-white, copyright-free illustrations. 96pp. 8¼ × 11. 24435-0 Pa. $3.95

CHANCERY CURSIVE STROKE BY STROKE, Arthur Baker. Instructions and illustrations for each stroke of each letter (upper and lower case) and numerals. 54 full-page plates. 64pp. 8¼ × 11. 24278-1 Pa. $2.50

THE ENJOYMENT AND USE OF COLOR, Walter Sargent. Color relationships, values, intensities; complementary colors, illumination, similar topics. Color in nature and art. 7 color plates, 29 illustrations. 274pp. 5⅜ × 8½. 20944-X Pa. $4.50

SCULPTURE PRINCIPLES AND PRACTICE, Louis Slobodkin. Step-by-step approach to clay, plaster, metals, stone; classical and modern. 253 drawings, photos. 255pp. 8⅛ × 11. 22960-2 Pa. $7.50

VICTORIAN FASHION PAPER DOLLS FROM HARPER'S BAZAR, 1867-1898, Theodore Menten. Four female dolls with 28 elegant high fashion costumes, printed in full color. 32pp. 9¼ × 12¼. 23453-3 Pa. $3.50

FLOPSY, MOPSY AND COTTONTAIL: A Little Book of Paper Dolls in Full Color, Susan LaBelle. Three dolls and 21 costumes (7 for each doll) show Peter Rabbit's siblings dressed for holidays, gardening, hiking, etc. Charming borders, captions. 48pp. 4¼ × 5½. 24376-1 Pa. $2.25

NATIONAL LEAGUE BASEBALL CARD CLASSICS, Bert Randolph Sugar. 83 big-leaguers from 1909-69 on facsimile cards. Hubbell, Dean, Spahn, Brock plus advertising, info, no duplications. Perforated, detachable. 16pp. 8¼ × 11. 24308-7 Pa. $2.95

THE LOGICAL APPROACH TO CHESS, Dr. Max Euwe, et al. First-rate text of comprehensive strategy, tactics, theory for the amateur. No gambits to memorize, just a clear, logical approach. 224pp. 5⅜ × 8½. 24353-2 Pa. $4.50

MAGICK IN THEORY AND PRACTICE, Aleister Crowley. The summation of the thought and practice of the century's most famous necromancer, long hard to find. Crowley's best book. 436pp. 5⅜ × 8½. (Available in U.S. only) 23295-6 Pa. $6.50

THE HAUNTED HOTEL, Wilkie Collins. Collins' last great tale; doom and destiny in a Venetian palace. Praised by T.S. Eliot. 127pp. 5⅜ × 8½. 24333-8 Pa. $3.00

ART DECO DISPLAY ALPHABETS, Dan X. Solo. Wide variety of bold yet elegant lettering in handsome Art Deco styles. 100 complete fonts, with numerals, punctuation, more. 104pp. 8⅛ × 11. 24372-9 Pa. $4.00

CALLIGRAPHIC ALPHABETS, Arthur Baker. Nearly 150 complete alphabets by outstanding contemporary. Stimulating ideas; useful source for unique effects. 154 plates. 157pp. 8⅜ × 11¼. 21045-6 Pa. $4.95

ARTHUR BAKER'S HISTORIC CALLIGRAPHIC ALPHABETS, Arthur Baker. From monumental capitals of first-century Rome to humanistic cursive of 16th century, 33 alphabets in fresh interpretations. 88 plates. 96pp. 9 × 12. 24054-1 Pa. $4.50

LETTIE LANE PAPER DOLLS, Sheila Young. Genteel turn-of-the-century family very popular then and now. 24 paper dolls. 16 plates in full color. 32pp. 9¼ × 12¼. 24089-4 Pa. $3.50

CATALOG OF DOVER BOOKS

HOW THE OTHER HALF LIVES, Jacob A. Riis. Journalistic record of filth, degradation, upward drive in New York immigrant slums, shops, around 1900. New edition includes 100 original Riis photos, monuments of early photography. 233pp. 10 × 7⅞.   22012-5 Pa. $7.95

CHINA AND ITS PEOPLE IN EARLY PHOTOGRAPHS, John Thomson. In 200 black-and-white photographs of exceptional quality photographic pioneer Thomson captures the mountains, dwellings, monuments and people of 19th-century China. 272pp. 9⅜ × 12¼.   24393-1 Pa. $12.95

GODEY COSTUME PLATES IN COLOR FOR DECOUPAGE AND FRAMING, edited by Eleanor Hasbrouk Rawlings. 24 full-color engravings depicting 19th-century Parisian haute couture. Printed on one side only. 56pp. 8¼ × 11.
23879-2 Pa. $3.95

ART NOUVEAU STAINED GLASS PATTERN BOOK, Ed Sibbett, Jr. 104 projects using well-known themes of Art Nouveau: swirling forms, florals, peacocks, and sensuous women. 60pp. 8¼ × 11.   23577-7 Pa. $3.50

QUICK AND EASY PATCHWORK ON THE SEWING MACHINE: Susan Aylsworth Murwin and Suzzy Payne. Instructions, diagrams show exactly how to machine sew 12 quilts. 48pp. of templates. 50 figures. 80pp. 8¼ × 11.
23770-2 Pa. $3.50

THE STANDARD BOOK OF QUILT MAKING AND COLLECTING, Marguerite Ickis. Full information, full-sized patterns for making 46 traditional quilts, also 150 other patterns. 483 illustrations. 273pp. 6⅞ × 9⅝. 20582-7 Pa. $5.95

LETTERING AND ALPHABETS, J. Albert Cavanagh. 85 complete alphabets lettered in various styles; instructions for spacing, roughs, brushwork. 121pp. 8¾ × 8.   20053-1 Pa. $3.75

LETTER FORMS: 110 COMPLETE ALPHABETS, Frederick Lambert. 110 sets of capital letters; 16 lower case alphabets; 70 sets of numbers and other symbols. 110pp. 8⅛ × 11.   22872-X Pa. $4.50

ORCHIDS AS HOUSE PLANTS, Rebecca Tyson Northen. Grow cattleyas and many other kinds of orchids—in a window, in a case, or under artificial light. 63 illustrations. 148pp. 5⅜ × 8½.   23261-1 Pa. $2.95

THE MUSHROOM HANDBOOK, Louis C.C. Krieger. Still the best popular handbook. Full descriptions of 259 species, extremely thorough text, poisons, folklore, etc. 32 color plates; 126 other illustrations. 560pp. 5⅜ × 8½.
21861-9 Pa. $8.50

THE DORÉ BIBLE ILLUSTRATIONS, Gustave Doré. All wonderful, detailed plates: Adam and Eve, Flood, Babylon, life of Jesus, etc. Brief King James text with each plate. 241 plates. 241pp. 9 × 12.   23004-X Pa. $8.95

THE BOOK OF KELLS: Selected Plates in Full Color, edited by Blanche Cirker. 32 full-page plates from greatest manuscript-icon of early Middle Ages. Fantastic, mysterious. Publisher's Note. Captions. 32pp. 9¾ × 12¼.   24345-1 Pa. $4.50

THE PERFECT WAGNERITE, George Bernard Shaw. Brilliant criticism of the Ring Cycle, with provocative interpretation of politics, economic theories behind Ring. 136pp. 5⅜ × 8½. (Available in U.S. only)   21707-8 Pa. $3.00

KEYBOARD WORKS FOR SOLO INSTRUMENTS, G.F. Handel. 35 neglected works from Handel's vast oeuvre, originally jotted down as improvisations. Includes Eight Great Suites, others. New sequence. 174pp. 9⅜ × 12¼.

24338-9 Pa. $7.50

AMERICAN LEAGUE BASEBALL CARD CLASSICS, Bert Randolph Sugar. 82 stars from 1900s to 60s on facsimile cards. Ruth, Cobb, Mantle, Williams, plus advertising, info, no duplications. Perforated, detachable. 16pp. 8¼ × 11.

24286-2 Pa. $2.95

A TREASURY OF CHARTED DESIGNS FOR NEEDLEWORKERS, Georgia Gorham and Jeanne Warth. 141 charted designs: owl, cat with yarn, tulips, piano, spinning wheel, covered bridge, Victorian house and many others. 48pp. 8¼ × 11.

23558-0 Pa. $1.95

DANISH FLORAL CHARTED DESIGNS, Gerda Bengtsson. Exquisite collection of over 40 different florals: anemone, Iceland poppy, wild fruit, pansies, many others. 45 illustrations. 48pp. 8¼ × 11.

23957-8 Pa. $1.75

OLD PHILADELPHIA IN EARLY PHOTOGRAPHS 1839-1914, Robert F. Looney. 215 photographs: panoramas, street scenes, landmarks, President-elect Lincoln's visit, 1876 Centennial Exposition, much more. 230pp. 8⅜ × 11¾.

23345-6 Pa. $9.95

PRELUDE TO MATHEMATICS, W.W. Sawyer. Noted mathematician's lively, stimulating account of non-Euclidean geometry, matrices, determinants, group theory, other topics. Emphasis on novel, striking aspects. 224pp. 5⅜ × 8½.

24401-6 Pa. $4.5(

ADVENTURES WITH A MICROSCOPE, Richard Headstrom. 59 adventur with clothing fibers, protozoa, ferns and lichens, roots and leaves, much more. illustrations. 232pp. 5⅜ × 8½.

23471-1 Pa. $

IDENTIFYING ANIMAL TRACKS: MAMMALS, BIRDS, AND OT' ANIMALS OF THE EASTERN UNITED STATES, Richard Headstror hunters, naturalists, scouts, nature-lovers. Diagrams of tracks, tips on ic cation. 128pp. 5⅜ × 8.

24442-3 P

VICTORIAN FASHIONS AND COSTUMES FROM HARPER'S BAZA 1898, edited by Stella Blum. Day costumes, evening wear, sports clothes, sh other accessories in over 1,000 detailed engravings. 320pp. 9⅜ × 12¼.

22990-

EVERYDAY FASHIONS OF THE TWENTIES AS PICTURED IN S OTHER CATALOGS, edited by Stella Blum. Actual dress of Twenties, with text by Stella Blum. Over 750 illustrations, captions.

241

HALL OF FAME BASEBALL CARDS, edited by Bert Randolph Su Ted Williams, Lou Gehrig, and many other Hall of Fame greats detachable reprints of early baseball cards. No duplication of c Baseball Cards. 16pp. 8¼ × 11.

THE ART OF HAND LETTERING, Helm Wotzkow. Cours Roman, Gothic, Italic, Block, Script. Tools, proportions, op dual variation. Very quality conscious. Hundreds of specime

THE RIME OF THE ANCIENT MARINER, Gustave Doré, S.T. Coleridge. Doré's finest work, 34 plates capture moods, subtleties of poem. Full text. 77pp. 9¼ × 12. 22305-1 Pa. $4.95

SONGS OF INNOCENCE, William Blake. The first and most popular of Blake's famous "Illuminated Books," in a facsimile edition reproducing all 31 brightly colored plates. Additional printed text of each poem. 64pp. 5¼ × 7.
22764-2 Pa. $3.00

AN INTRODUCTION TO INFORMATION THEORY, J.R. Pierce. Second (1980) edition of most impressive non-technical account available. Encoding, entropy, noisy channel, related areas, etc. 320pp. 5⅜ × 8½. 24061-4 Pa. $4.95

THE DIVINE PROPORTION: A STUDY IN MATHEMATICAL BEAUTY, H.E. Huntley. "Divine proportion" or "golden ratio" in poetry, Pascal's triangle, philosophy, psychology, music, mathematical figures, etc. Excellent bridge between science and art. 58 figures. 185pp. 5⅜ × 8½. 22254-3 Pa. $3.95

THE DOVER NEW YORK WALKING GUIDE: From the Battery to Wall Street, Mary J. Shapiro. Superb inexpensive guide to historic buildings and locales in lower Manhattan: Trinity Church, Bowling Green, more. Complete Text; maps. 36 illustrations. 48pp. 3⅞ × 9¼. 24225-0 Pa. $2.50

NEW YORK THEN AND NOW, Edward B. Watson, Edmund V. Gillon, Jr. 83 important Manhattan sites: on facing pages early photographs (1875-1925) and 1976 photos by Gillon. 172 illustrations. 171pp. 9¼ × 10. 23361-8 Pa. $7.95

HISTORIC COSTUME IN PICTURES, Braun & Schneider. Over 1450 costumed figures from dawn of civilization to end of 19th century. English captions. 125 plates. 256pp. 8⅜ × 11¼. 23150-X Pa. $7.50

VICTORIAN AND EDWARDIAN FASHION: A Photographic Survey, Alison Gernsheim. First fashion history completely illustrated by contemporary photographs. Full text plus 235 photos, 1840-1914, in which many celebrities appear. 240pp. 6½ × 9¼. 24205-6 Pa. $6.00

CHARTED CHRISTMAS DESIGNS FOR COUNTED CROSS-STITCH AND OTHER NEEDLECRAFTS, Lindberg Press. Charted designs for 45 beautiful needlecraft projects with many yuletide and wintertime motifs. 48pp. 8¼ × 11.
24356-7 Pa. $1.95

101 FOLK DESIGNS FOR COUNTED CROSS-STITCH AND OTHER NEEDLE-CRAFTS, Carter Houck. 101 authentic charted folk designs in a wide array of lovely representations with many suggestions for effective use. 48pp. 8¼ × 11.
24369-9 Pa. $2.25

FIVE ACRES AND INDEPENDENCE, Maurice G. Kains. Great back-to-the-land classic explains basics of self-sufficient farming. The one book to get. 95 illustrations. 397pp. 5⅜ × 8½. 20974-1 Pa. $4.95

A MODERN HERBAL, Margaret Grieve. Much the fullest, most exact, most useful compilation of herbal material. Gigantic alphabetical encyclopedia, from aconite to zedoary, gives botanical information, medical properties, folklore, economic uses, and much else. Indispensable to serious reader. 161 illustrations. 888pp. 6½ × 9¼. (Available in U.S. only) 22798-7, 22799-5 Pa., Two-vol. set $16.45

DECORATIVE NAPKIN FOLDING FOR BEGINNERS, Lillian Oppenheimer and Natalie Epstein. 22 different napkin folds in the shape of a heart, clown's hat, love knot, etc. 63 drawings. 48pp. 8¼ × 11. 23797-4 Pa. $1.95

DECORATIVE LABELS FOR HOME CANNING, PRESERVING, AND OTHER HOUSEHOLD AND GIFT USES, Theodore Menten. 128 gummed, perforated labels, beautifully printed in 2 colors. 12 versions. Adhere to metal, glass, wood, ceramics. 24pp. 8¼ × 11. 23219-0 Pa. $2.95

EARLY AMERICAN STENCILS ON WALLS AND FURNITURE, Janet Waring. Thorough coverage of 19th-century folk art: techniques, artifacts, surviving specimens. 166 illustrations, 7 in color. 147pp. of text. 7⅞ × 10¾. 21906-2 Pa. $9.95

AMERICAN ANTIQUE WEATHERVANES, A.B. & W.T. Westervelt. Extensively illustrated 1883 catalog exhibiting over 550 copper weathervanes and finials. Excellent primary source by one of the principal manufacturers. 104pp. 6⅛ × 9¼.
24396-6 Pa. $3.95

ART STUDENTS' ANATOMY, Edmond J. Farris. Long favorite in art schools. Basic elements, common positions, actions. Full text, 158 illustrations. 159pp. 5⅜ × 8½. 20744-7 Pa. $3.95

BRIDGMAN'S LIFE DRAWING, George B. Bridgman. More than 500 drawings and text teach you to abstract the body into its major masses. Also specific areas of anatomy. 192pp. 6½ × 9¼. (EA) 22710-3 Pa. $4.50

COMPLETE PRELUDES AND ETUDES FOR SOLO PIANO, Frederic Chopin. All 26 Preludes, all 27 Etudes by greatest composer of piano music. Authoritative Paderewski edition. 224pp. 9 × 12. (Available in U.S. only) 24052-5 Pa. $7.50

PIANO MUSIC 1888-1905, Claude Debussy. Deux Arabesques, Suite Bergamesque, Masques, 1st series of Images, etc. 9 others, in corrected editions. 175pp. 9⅜ × 12¼.
(ECE) 22771-5 Pa. $5.95

TEDDY BEAR IRON-ON TRANSFER PATTERNS, Ted Menten. 80 iron-on transfer patterns of male and female Teddys in a wide variety of activities, poses, sizes. 48pp. 8¼ × 11. 24596-9 Pa. $2.25

A PICTURE HISTORY OF THE BROOKLYN BRIDGE, M.J. Shapiro. Profusely illustrated account of greatest engineering achievement of 19th century. 167 rare photos & engravings recall construction, human drama. Extensive, detailed text. 122pp. 8¼ × 11. 24403-2 Pa. $7.95

NEW YORK IN THE THIRTIES, Berenice Abbott. Noted photographer's fascinating study shows new buildings that have become famous and old sights that have disappeared forever. 97 photographs. 97pp. 11⅜ × 10. 22967-X Pa. $6.50

MATHEMATICAL TABLES AND FORMULAS, Robert D. Carmichael and Edwin R. Smith. Logarithms, sines, tangents, trig functions, powers, roots, reciprocals, exponential and hyperbolic functions, formulas and theorems. 269pp. 5⅜ × 8½. 60111-0 Pa. $3.75

HANDBOOK OF MATHEMATICAL FUNCTIONS WITH FORMULAS, GRAPHS, AND MATHEMATICAL TABLES, edited by Milton Abramowitz and Irene A. Stegun. Vast compendium: 29 sets of tables, some to as high as 20 places. 1,046pp. 8 × 10½. 61272-4 Pa. $19.95

REASON IN ART, George Santayana. Renowned philosopher's provocative, seminal treatment of basis of art in instinct and experience. Volume Four of *The Life of Reason.* 230pp. 5⅜ × 8. 24358-3 Pa. $4.50

LANGUAGE, TRUTH AND LOGIC, Alfred J. Ayer. Famous, clear introduction to Vienna, Cambridge schools of Logical Positivism. Role of philosophy, elimination of metaphysics, nature of analysis, etc. 160pp. 5⅜ × 8½. (USCO) 20010-8 Pa. $2.75

BASIC ELECTRONICS, U.S. Bureau of Naval Personnel. Electron tubes, circuits, antennas, AM, FM, and CW transmission and receiving, etc. 560 illustrations. 567pp. 6½ × 9¼. 21076-6 Pa. $8.95

THE ART DECO STYLE, edited by Theodore Menten. Furniture, jewelry, metalwork, ceramics, fabrics, lighting fixtures, interior decors, exteriors, graphics from pure French sources. Over 400 photographs. 183pp. 8⅜ × 11¼. 22824-X Pa. $6.95

THE FOUR BOOKS OF ARCHITECTURE, Andrea Palladio. 16th-century classic covers classical architectural remains, Renaissance revivals, classical orders, etc. 1738 Ware English edition. 216 plates. 110pp. of text. 9½ × 12¾. 21308-0 Pa. $11.50

THE WIT AND HUMOR OF OSCAR WILDE, edited by Alvin Redman. More than 1000 ripostes, paradoxes, wisecracks: Work is the curse of the drinking classes, I can resist everything except temptations, etc. 258pp. 5⅜ × 8½. (USCO) 20602-5 Pa. $3.50

THE DEVIL'S DICTIONARY, Ambrose Bierce. Barbed, bitter, brilliant witticisms in the form of a dictionary. Best, most ferocious satire America has produced. 145pp. 5⅜ × 8½. 20487-1 Pa. $2.50

ERTÉ'S FASHION DESIGNS, Erté. 210 black-and-white inventions from *Harper's Bazar,* 1918-32, plus 8pp. full-color covers. Captions. 88pp. 9 × 12. 24203-X Pa. $6.50

ERTÉ GRAPHICS, Erté. Collection of striking color graphics: *Seasons, Alphabet, Numerals, Aces* and *Precious Stones.* 50 plates, including 4 on covers. 48pp. 9⅜ × 12¼. 23580-7 Pa. $6.95

PAPER FOLDING FOR BEGINNERS, William D. Murray and Francis J. Rigney. Clearest book for making origami sail boats, roosters, frogs that move legs, etc. 40 projects. More than 275 illustrations. 94pp. 5⅜ × 8½. 20713-7 Pa. $2.25

ORIGAMI FOR THE ENTHUSIAST, John Montroll. Fish, ostrich, peacock, squirrel, rhinoceros, Pegasus, 19 other intricate subjects. Instructions. Diagrams. 128pp. 9 × 12. 23799-0 Pa. $4.95

CROCHETING NOVELTY POT HOLDERS, edited by Linda Macho. 64 useful, whimsical pot holders feature kitchen themes, animals, flowers, other novelties. Surprisingly easy to crochet. Complete instructions. 48pp. 8¼ × 11. 24296-X Pa. $1.95

CROCHETING DOILIES, edited by Rita Weiss. Irish Crochet, Jewel, Star Wheel, Vanity Fair and more. Also luncheon and console sets, runners and centerpieces. 51 illustrations. 48pp. 8¼ × 11. 23424-X Pa. $2.00

YUCATAN BEFORE AND AFTER THE CONQUEST, Diego de Landa. Only significant account of Yucatan written in the early post-Conquest era. Translated by William Gates. Over 120 illustrations. 162pp. 5⅜ × 8½. 23622-6 Pa. $3.50

ORNATE PICTORIAL CALLIGRAPHY, E.A. Lupfer. Complete instructions, over 150 examples help you create magnificent "flourishes" from which beautiful animals and objects gracefully emerge. 8⅛ × 11. 21957-7 Pa. $2.95

DOLLY DINGLE PAPER DOLLS, Grace Drayton. Cute chubby children by same artist who did Campbell Kids. Rare plates from 1910s. 30 paper dolls and over 100 outfits reproduced in full color. 32pp. 9¼ × 12¼. 23711-7 Pa. $3.50

CURIOUS GEORGE PAPER DOLLS IN FULL COLOR, H. A. Rey, Kathy Allert. Naughty little monkey-hero of children's books in two doll figures, plus 48 full-color costumes: pirate, Indian chief, fireman, more. 32pp. 9¼ × 12¼. 24386-9 Pa. $3.50

GERMAN: HOW TO SPEAK AND WRITE IT, Joseph Rosenberg. Like *French, How to Speak and Write It.* Very rich modern course, with a wealth of pictorial material. 330 illustrations. 384pp. 5⅜ × 8½. (USUKO) 20271-2 Pa. $4.75

CATS AND KITTENS: 24 Ready-to-Mail Color Photo Postcards, D. Holby. Handsome collection; feline in a variety of adorable poses. Identifications. 12pp. on postcard stock. 8¼ × 11. 24469-5 Pa. $2.95

MARILYN MONROE PAPER DOLLS, Tom Tierney. 31 full-color designs on heavy stock, from *The Asphalt Jungle, Gentlemen Prefer Blondes,* 22 others. 1 doll. 16 plates. 32pp. 9⅜ × 12¼. 23769-9 Pa. $3.50

FUNDAMENTALS OF LAYOUT, F.H. Wills. All phases of layout design discussed and illustrated in 121 illustrations. Indispensable as student's text or handbook for professional. 124pp. 8⅛ × 11. 21279-3 Pa. $4.50

FANTASTIC SUPER STICKERS, Ed Sibbett, Jr. 75 colorful pressure-sensitive stickers. Peel off and place for a touch of pizzazz: clowns, penguins, teddy bears, etc. Full color. 16pp. 8¼ × 11. 24471-7 Pa. $2.95

LABELS FOR ALL OCCASIONS, Ed Sibbett, Jr. 6 labels each of 16 different designs—baroque, art nouveau, art deco, Pennsylvania Dutch, etc.—in full color. 24pp. 8¼ × 11. 23688-9 Pa. $2.95

HOW TO CALCULATE QUICKLY: RAPID METHODS IN BASIC MATHEMATICS, Henry Sticker. Addition, subtraction, multiplication, division, checks, etc. More than 8000 problems, solutions. 185pp. 5 × 7¼. 20295-X Pa. $2.95

THE CAT COLORING BOOK, Karen Baldauski. Handsome, realistic renderings of 40 splendid felines, from American shorthair to exotic types. 44 plates. Captions. 48pp. 8¼ × 11. 24011-8 Pa. $2.25

THE TALE OF PETER RABBIT, Beatrix Potter. The inimitable Peter's terrifying adventure in Mr. McGregor's garden, with all 27 wonderful, full-color Potter illustrations. 55pp. 4¼ × 5½. (Available in U.S. only) 22827-4 Pa. $1.60

BASIC ELECTRICITY, U.S. Bureau of Naval Personnel. Batteries, circuits, conductors, AC and DC, inductance and capacitance, generators, motors, transformers, amplifiers, etc. 349 illustrations. 448pp. 6½ × 9¼. 20973-3 Pa. $7.95

SOURCE BOOK OF MEDICAL HISTORY, edited by Logan Clendening, M.D. Original accounts ranging from Ancient Egypt and Greece to discovery of X-rays: Galen, Pasteur, Lavoisier, Harvey, Parkinson, others. 685pp. 5⅜ × 8½.

20621-1 Pa. $10.95

THE ROSE AND THE KEY, J.S. Lefanu. Superb mystery novel from Irish master. Dark doings among an ancient and aristocratic English family. Well-drawn characters; capital suspense. Introduction by N. Donaldson. 448pp. 5⅜ × 8½.

24377-X Pa. $6.95

SOUTH WIND, Norman Douglas. Witty, elegant novel of ideas set on languorous Mediterranean island of Nepenthe. Elegant prose, glittering epigrams, mordant satire. 1917 masterpiece. 416pp. 5⅜ × 8½. (Available in U.S. only)

24361-3 Pa. $5.95

RUSSELL'S CIVIL WAR PHOTOGRAPHS, Capt. A.J. Russell. 116 rare Civil War Photos: Bull Run, Virginia campaigns, bridges, railroads, Richmond, Lincoln's funeral car. Many never seen before. Captions. 128pp. 9⅜ × 12¼.

24283-8 Pa. $6.95

PHOTOGRAPHS BY MAN RAY: 105 Works, 1920-1934. Nudes, still lifes, landscapes, women's faces, celebrity portraits (Dali, Matisse, Picasso, others), rayographs. Reprinted from rare gravure edition. 128pp. 9⅜ × 12¼. (Available in U.S. only)

23842-3 Pa. $6.95

STAR NAMES: THEIR LORE AND MEANING, Richard H. Allen. Star names, the zodiac, constellations: folklore and literature associated with heavens. The basic book of its field, fascinating reading. 563pp. 5⅜ × 8½.

21079-0 Pa. $7.95

BURNHAM'S CELESTIAL HANDBOOK, Robert Burnham, Jr. Thorough guide to the stars beyond our solar system. Exhaustive treatment. Alphabetical by constellation: Andromeda to Cetus in Vol. 1; Chamaeleon to Orion in Vol. 2; and Pavo to Vulpecula in Vol. 3. Hundreds of illustrations. Index in Vol. 3. 2000pp. 6⅛ × 9¼.

23567-X, 23568-8, 23673-0 Pa. Three-vol. set $36.85

THE ART NOUVEAU STYLE BOOK OF ALPHONSE MUCHA, Alphonse Mucha. All 72 plates from *Documents Decoratifs* in original color. Stunning, essential work of Art Nouveau. 80pp. 9⅜ × 12¼.

24044-4 Pa. $7.95

DESIGNS BY ERTE; FASHION DRAWINGS AND ILLUSTRATIONS FROM "HARPER'S BAZAR," Erte. 310 fabulous line drawings and 14 *Harper's Bazar* covers, 8 in full color. Erte's exotic temptresses with tassels, fur muffs, long trains, coifs, more. 129pp. 9⅜ × 12¼.

23397-9 Pa. $6.95

HISTORY OF STRENGTH OF MATERIALS, Stephen P. Timoshenko. Excellent historical survey of the strength of materials with many references to the theories of elasticity and structure. 245 figures. 452pp. 5⅜ × 8½. 61187-6 Pa. $8.95

Prices subject to change without notice.

Available at your book dealer or write for free catalog to Dept. GI, Dover Publications, Inc., 31 East 2nd St. Mineola, N.Y. 11501. Dover publishes more than 175 books each year on science, elementary and advanced mathematics, biology, music, art, literary history, social sciences and other areas.